Signalling by Inositides

W9-CGK-487

The Practical Approach Series

SERIES EDITOR

B. D. HAMES
Department of Biochemistry and Molecular Biology
University of Leeds, Leeds LS2 9JT, UK

★ **indicates new and forthcoming titles**

Signalling by Inositides

A Practical Approach

Edited by
STEPHEN SHEARS
National Institute of Environmental Health Sciences,
North Carolina, USA

OXFORD UNIVERSITY PRESS
Oxford New York Tokyo

Oxford University Press, Great Clarendon Street, Oxford OX2 6DP

Oxford   New York
Athens   Auckland   Bangkok   Bogota   Bombay   Buenos Aires
Calcutta   Cape Town   Dar es Salaam   Delhi   Florence   Hong Kong
Istanbul   Karachi   Kuala Lumpur   Madras   Madrid   Melbourne
Mexico City   Nairobi   Paris   Singapore   Taipei   Tokyo   Toronto
and associated companies in
Berlin   Ibadan

Oxford is a trade mark of Oxford University Press

Published in the United States
by Oxford University Press Inc., New York

© Oxford University Press, 1997

All rights reserved. No part of this publication may be
reproduced, stored in a retrieval system, or transmitted, in any
form or by any means, without the prior permission in writing of Oxford
University Press. Within the UK, exceptions are allowed in respect of any
fair dealing for the purpose of research or private study, or criticism or
review, as permitted under the Copyright, Designs and Patents Act, 1988, or
in the case of reprographic reproduction in accordance with the terms of
licences issued by the Copyright Licensing Agency. Enquiries concerning
reproduction outside those terms and in other countries should be sent to
the Rights Department, Oxford University Press, at the address above.

This book is sold subject to the condition that it shall not,
by way of trade or otherwise, be lent, re-sold, hired out, or otherwise
circulated without the publisher's prior consent in any form of binding
or cover other than that in which it is published and without a similar
condition including this condition being imposed
on the subsequent purchaser.

Users of books in the Practical Approach Series are advised that prudent
laboratory safety procedures should be followed at all times. Oxford
University Press makes no representation, express or implied, in respect of
the accuracy of the material set forth in books in this series and cannot
accept any legal responsibility or liability for any errors or omissions
that may be made.

A catalogue record for this book is available from the British Library

Library of Congress Cataloging in Publication Data
(Data available)
ISBN 0 19 963639 7 (Hbk)
ISBN 0 19 963638 9 (Pbk)

Typeset by Footnote Graphics, Warminster, Wilts
Printed in Great Britain by Information Press, Ltd, Eynsham, Oxon.

Preface

One goal of this book has been to gather together details of some of the most fundamental and commonly used methodologies in inositide research (that is, studies into the metabolism and functions of inositol phosphates and inositol lipids). A second goal was to secure contributions from a selection of the leading international signal transduction laboratories which have been actively involved in the development and/or refinement of these techniques. The extent to which these aims have been met is due in no small part to the enthusiasm of the authors, their attention to detail, and an adherence to a strict time-table during this book's production (with relatively little badgering on my part!).

As with other volumes of this Practical Approach series, the end result had to be sufficiently compact that it could qualify as an economically feasible laboratory manual. In any case, this book could not hope to be comprehensive; it would be a truly encyclopedic undertaking to embrace the many areas of biochemistry, physiology, and molecular biology that have been infiltrated by inositide research. Instead, I have tried to concentrate on some of the basics. In addition, certain topics were excluded because they had fairly recently been covered elsewhere. For example, the techniques used to study glycosyl-phosphatidylinositol anchors were reviewed in detail (over 150 pages) by a 1995 volume of Methods in Enzymology (vol 250, ed. P. J. Casey and J. Buss, Academic Press, New York).

I also saw this volume as an opportunity to gather together a list of non-commercial sources of synthetic inositide derivatives that have been prepared by a number of laboratories from around the world. I hope that a wider appreciation of their availability will result in an increased application of these compounds to some important problems. Under appropriate management—a scientific society perhaps—this particular project could evolve into a page on the internet so that it could be expanded and updated.

I would like to thank all those at Oxford University Press for their encouragement and assistance, and for minimizing the production time for this book.

Finally, I must acknowledge the support and patience of my wife, Jane, during the many times work on this project has conflicted with our other commitments.

Research Triangle Park   S. S.
May 1997

Contents

Contents

3. Mass assay of inositol and its use to assay inositol polyphosphates

Tamas Balla

4. Expression, purification and interfacial kinetic analysis of phospholipase C-γ1

Gwenith A. Jones and Debra A. Horstman

Contents

12. Phosphoinositide 3-kinase
203

Sebastian Pons, Deborah J. Burks, and Morris F. White

A1. List of suppliers
215

A2. List of suppliers of inositol derivatives
217

Index
227

Contributors

TAMAS BALLA
Endocrinology and Reproduction Research Branch, National Institute of Child Health and Human Development, National Institutes of Health, Bethesda, MD 20892, USA.

MICHAEL E. BEMBENEK
E. I. Du Pont Nemours & Co. Inc., NEN Medical Products Department, Boston, Massachusetts 02118, USA.

CHARLES A. BREARLEY
Department of Plant Sciences, University of Cambridge, Downing St., Cambridge CB2 3EA, UK.

DEBORAH J. BURKS
Joslin Diabetes Center, Harvard Medical School, One Joslin Place, Boston, MA 02215, USA.

R. A. JOHN CHALLISS
Department of Cell Physiology and Pharmacology, University of Leicester, PO Box 138, Medical Sciences Building, University Road, Leicester LE1 9HN, UK.

GYÖRGY DORMÁN
The State University of New York, Department of Chemistry, Stony Brook, New York 11794, USA.

YASUFUMI EMORI
Department of Biophysics and Biochemistry, Faculty of Science, University of Tokyo, Tokyo 113, Japan.

THERESA M. FILTZ
Department of Pharmacology, Faculty Laboratory Office Building, CB# 7365, University of North Carolina School of Medicine, Chapel Hill, N.C. 27599-7265, USA.

LATANYA P. HAMMONDS-ODIE
Departments of Neurobiology and Cell Biology, University of Alabama at Birmingham, Rm. 576 CIRC, 1719 Sixth Avenue South, Birmingham, AL 35294, USA.

DAVID E. HANKE
Department of Plant Sciences, University of Cambridge, Downing St., Cambridge CB2 3EA, UK.

Contributors

T. KENDALL HARDEN
Department of Pharmacology, Faculty Laboratory Office Building, CB# 7365, University of North Carolina School of Medicine, Chapel Hill, N.C. 27599-7265, USA.

YOSHIMI HOMMA
Department of Biomolecular Sciences, Fukushima Medical College, Fukushima 960–12, Japan

DEBRA A. HORSTMAN
Department of Biochemistry, Vanderbilit University School of Medicine, Nashville, Tennessee 37232, USA.

TREVOR R. JACKSON
Laboratory of Molecular Signalling, Babraham Institute, Department of Zoology, University of Cambridge, Cambridge CB2 3ES, UK.

GWENITH A. JONES
Department of Pharmacology, University of Virginia, Charlottesville Virginia 22908, USA.

ADAM I. KAPLIN
Johns Hopkins University School of Medicine, Departments of Neuroscience, Pharmacology and Molecular Sciences, and Psychiatry, Baltimore, Maryland 21205, USA.

KATSUHIKO MIKOSHIBA
Department of Molecular Neurobiology, The Institute of Medical Science, The University of Tokyo, Minato-ku, Tokyo 108, Japan.

JOHN D. OLSZEWSKI
The State University of New York, Department of Chemistry, Stony Brook, New York 11794, USA.

ANDREW PATERSON
Department of Pharmacology, Faculty Laboratory Office Building, CB# 7365, University of North Carolina School of Medicine, Chapel Hill, N.C. 27599-7265, USA.

SEBASTIAN PONS
Joslin Diabetes Center, Harvard Medical School, One Joslin Place, Boston, MA 02215, USA.

GLENN D. PRESTWICH
Department of Medicinal Chemistry, University of Utah, Salt Lake City Utah, UT 84112, USA.

STEPHEN B. SHEARS
Laboratory of Cellular and Molecular Pharmacology, National Institute of Environmental Health Sciences, Research Triangle Park, NC 27709, USA.

Contributors

SOLOMON H. SNYDER
Johns Hopkins University School of Medicine, Departments of Neuroscience, Pharmacology and Molecular Sciences, and Psychiatry, Baltimore, Maryland 21205, USA.

ANNE B. THEIBERT
Departments of Neurobiology and Cell Biology, University of Alabama at Birmingham, Rm. 576 CIRC, 1719 Sixth Avenue South, Birmingham, AL 35294, USA.

JEROEN VAN DER KAAY
Department of Biochemistry, University of Groningen, Nijenborgh 4, 9747 AG Groningen, The Netherlands.

PETER J. M. VAN HAASTERT
Department of Biochemistry, University of Groningen, Nijenborgh 4, 9747 AG Groningen, The Netherlands.

SUSAN M. VOGLMAIER
Johns Hopkins University School of Medicine, Departments of Neuroscience, Pharmacology and Molecular Sciences, and Psychiatry, Baltimore, Maryland 21205, USA.

MORRIS F. WHITE
Joslin Diabetes Center, Harvard Medical School, One Joslin Place, Boston, MA 02215, USA.

Abbreviations

ADP	adenosine diphosphate
AMP	adenosine monophosphate
ASA	arylaside
ATP	adenosine triphosphate
BSA	bovine serum albumin
BZDC	benzophenone
CDP	cytosine diphosphate
CDP-DG	cytosine diphosphate diacylglycerol
CDTA	trans-1,2-diaminocyclohexane-N,N,N,N'-tetraacetic acid
CMP	cytosine monophosphate
CMP-PtdOH	cytosine monophosphate phosphatidate
CTP	cytosine triphosphate
DAG	diacylglycerol
DTT	dithiothreitol
EDTA	ethylenediaminetetraacetic acid
EGF	epidermal growth factor
EGTA	ethyleneglycol-*bis*-(β-aminoethyl ether)-N,N,N',N'-tetraacetic acid
ELISA	enzyme-linked immunosorbant assay
ER	endoplasmic reticulum
FGF	fibroblast growth factor
FPLC	fast protein liquid chromatography
GAP	GTPase activating protein
G-protein	guanine nucleotide regulatory protein
GroP	glycerol phosphate
GroPGro	glycerol phosphoglycerol
GroPIns3P	glycerophosphoinositol 3-monophosphate
GroPIns4P	glycerophosphoinositol 4-monophosphate
GroPIns	glycerophosphoinositol
GroPIns$(3,4)P_2$	glycerophosphoinositol 3,4-bisphosphate
GroPIns$(3,4,5)P_3$	glycerophosphoinositol 3,4,5-trisphosphate
GroPIns$(4,5)P_2$	glycerophosphoinositol 4,5-bisphosphate
GroPP	glycerol pyrophosphate
GST	glutathione S-transferase
GTP	guanosine triphosphate
HPLC	high performance liquid chromatography
IDH	*myo*-inositol dehydrogenase
Ins$(1,3)P_2$	inositol 1,3-bisphosphate
Ins$(1,3,4)P_3$	inositol 1,3,4-trisphosphate
Ins$(1,3,4,5)P_4$	inositol 1,3,4,5-tetrakisphosphate

Ins(1,4)P_2	inositol 1,4-bisphosphate
Ins(1,4,5)P_3	inositol 1,4,5-trisphosphate
Ins(3,4)P_2	inositol 3,4-bisphosphate
InsP_5	inositol pentakisphosphate
InsP_6	inositol hexakisphosphate
InsP_n	inositol polyphosphate
MOI	mutiplicity of infection
MW	molecular weight
MWCO	molecular weight cut-off
NAD	nicotinamide–adenine dinucleotide
PCR	polymerase chain reaction
PDGF	platelet-derived growth factor
PEG	polyethyleneglycol
PEI	polyethyleneimine
p.f.u	plaque forming units
PH	pleckstrin homology
PI 3-kinase	phosphoinositide 3-kinase
P_i	inorganic phosphate
PKB	Akt kinase
PKC	protein kinase C
PLC	inositol lipid-directed phospholipase C
PLC-γ1	phospholipase C-γ1
PMSF	phenylmethylsulfonyl fluoride
PP	pyrophosphate
PRK	protein kinase C-related kinase
PtdIns3P	phosphatidylinositol 3-monophosphate
PtdIns4P	phosphatidylinositol 4-monophosphate
PtdIns	phosphatidylinositol
PtdIns 3-kinase	phosphatidylinositol 3-kinase
PtdIns 4-kinase	phosphatidylinositol 4-kinase
PtdIns(3,4)P_2	phosphatidylinositol 3,4-bisphosphate
PtdIns(3,4,5)P_3	phosphatidylinositol 3,4,5-trisphosphate
PtdIns(4,5)P_2	phosphatidylinositol 4,5-bisphosphate
PtdOH	phosphatidic acid
PtdInsP_n	inositol lipid
SAX	strong anion exchange
SDS-PAGE	SDS polyacrylamide gel electrophoresis
Sf9	*Spodoptera frugiperda*
TCA	trichloroacetic acid
TEAB	triethylamine bicarbonate
TLC	thin-layer chromatography
TSH	thyroid stimulating hormone

1

Assaying inositol phospholipid turnover in plant cells

CHARLES A. BREARLEY and DAVID E. HANKE

1. Introduction

The inexorable growth in the complexity of inositol phospholipid and inositol phosphate science, principally in the animal field but also in yeast, has served to emphasize our lack of understanding of even the fundamentals in plant cells. The fragmented nature of plant inositol phosphate research is a primary problem (1). The lack of the plant equivalent of the avian or human erythrocyte, which features throughout this chapter as a model system in which virtually all the methods described here were developed or tested, holds back the efforts of many plant scientists. Thus, there is much to be gained from seeking a simple and, most importantly, rigorous description of the identities of the numerous inositol phosphates and inositol phospholipids present in plant tissues before proceeding to the rather more complicated task of investigating their function. The techniques detailed in this chapter afford a powerful strategy for identifying some of these components, investigating the fundamentals of inositol phospholipid turnover in plant cells and beyond this testing for the involvement of these components in the responses of plants to a variety of environmental, physiological, pathological, and developmental stimuli. We shall pay special attention to a group of inositol phospholipids, phosphorylated in the D-3 position, which form a cohort of novel signalling lipids. Recent work suggests that there are substantive differences in the metabolism of 3-phosphorylated lipids in plants, animals and yeast.

2. Structures of inositol phospholipids

We have limited this chapter to a discussion of inositol glycerophospholipids. Although inositol sphingolipids have been reported in plants much less is known of their structure or even function and this is reflected in considerable confusion in the nomenclature of these lipids. In contrast, the burgeoning interest over the last 20 years in inositol phospholipids and inositol phosphates and their roles in cell function has driven the development of a wide range of analytical techniques which have only recently been applied to plants.

The multiple possibilities for substitution of the inositol moiety of *myo*-inositol phosphates means that 63 non-cyclic stereoisomers are possible (not counting the recently discovered inositol pyrophosphates). Fortunately the range of inositol phospholipids is not so complex. Similarly, the added complication of the existence, not just in theory but *in vivo*, of enantiomeric forms of *myo*-inositol phosphates has not yet been realized for the lipids. The development of techniques for discriminating between stereoisomers and enantiomers of the numerous inositol phosphates and the application of these techniques to lipids has facilitated uncompromized analysis of their structure.

2.1 Metabolic relations of inositol phospholipids

The common structural unit of all phosphatidylinositols (*Figure 1*) is phosphatidyl-1-D-*myo*- inositol (PtdIns). The ultimate phospholipid precursor of PtdIns is phosphatidic acid (PtdOH). The route of *de novo* synthesis of PtdIns from PtdOH involves two steps (*Figure 2*). PtdOH condenses with CTP to form CDP-diacylglycerol (this product is more informatively called CMP-phosphatidate in older texts and CMP-PtdOH in the present text). The other product is pyrophosphate, comprising the β- and γ-phosphates of the CTP. The subsequent hydrolysis of pyrophosphate renders the formation of PtdIns irreversible. CMP-PtdOH condenses with *myo*-inositol producing PtdIns and CMP.

Although there are various potential routes to PtdOH in plants, it is generally considered that in animal cells the principal route *in vivo* is phosphorylation of diacylglycerol by diacylglycerol kinase. Thus the γ-phosphate of ATP is the direct metabolic precursor of the α-phosphate of PtdOH and hence of the diester phosphate of PtdIns. Deacylation of phosphatidylinositols yields glycerophosphoinositols, which on deglyceration yield inositol phosphates in which the configuration of substitution of the inositol moiety is maintained (*Figure 3*). Consequently, the diester phosphate of the different phosphatidylinositols is synonymous with the D-1 phosphate of the inositol phosphate derivatives of these lipids.

The inositol phospholipids contain a *myo*-inositol headgroup, i.e. all the hydroxyl groups are equatorial except for the axial group at C_2. *Myo*-inositol has a plane of symmetry between C_2 and C_5. Consequently, substitution in positions other than C_2 and C_5 generates the possibility of enantiomeric forms of substituted inositols. Since enantiomeric forms of phosphatidylinositols have not yet been detected, we need only concern ourselves with the distinction between stereoisomers.

Although the function of inositol phospholipids is beyond the remit of this text an outline of the metabolic relations of the various inositol phospholipids is of value. The techniques described below have facilitated the description of

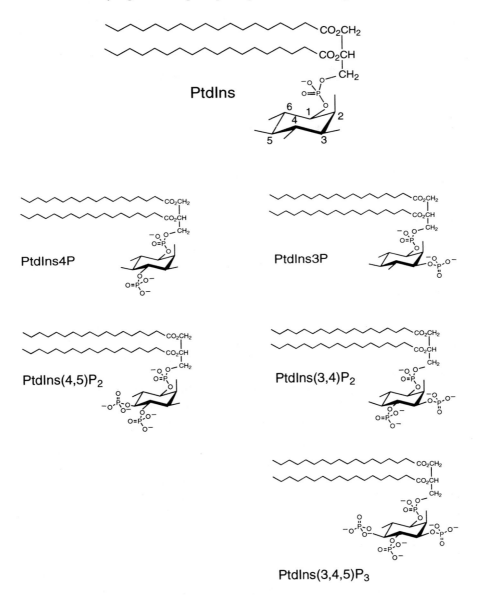

Figure 1. Structures of inositol phospholipids. PtdIns, phosphatidylinositol (phosphomonoester phosphates identified relative to the D-1 phosphodiester phosphate of PtdIns). PtdIns3P, phosphatidylinositol 3-monophosphate; PtdIn4P, phosphatidylinositol 4-monophosphate; PtdIns(3,4)P_2, phosphatidylinositol 3,4-bisphosphate; PtdIns(4,5)P_2, phosphatidylinositol 4,5-bisphosphate; PtdIns(3,4,5)P_3, phosphatidylinositol 3,4,5-trisphosphate. PtdIns(3,4)P_2 and PtdIns(3,4,5)P_3 have not been identified in yeast. PtdIns(3,4,5)P_3 has not been identified in plants.

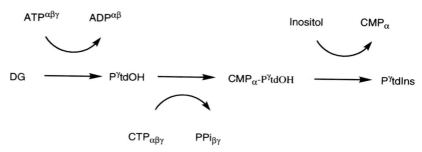

Figure 2. Route of synthesis of phosphatidylinositol. The origins of the individual phosphate moieties, α, β, γ, of intermediates in the metabolic sequence are indicated. CMP, cytosine monophosphate; CMP-PtdOH, cytosine monophosphate phosphatidate; CTP, cytosine trisphosphate; DG, diacylglycerol; PP$_i$, pyrophosphate; PtdOH, phosphatidic acid; PtdIns, phosphatidylinositol.

the routes of synthesis and turnover of the different inositol phospholipids. Much supporting evidence has been provided by *in vitro* approaches. However, there is some debate in the literature regarding the routes of synthesis of the different lipids, though this is restricted to 3-phosphorylated lipids.

It is generally accepted (2, 3, 4) that in animal cells PtdIns(3,4,5)P_3 is synthesized by 3- phosphorylation of PtdIns(4,5)P_2 (*Figure 4*). PtdIns(4,5)P_2 3-kinase is acutely sensitive to activation in both a G-protein (2, 5) and receptor tyrosine kinase (2, 6) dependent manner. Animal cells possess phosphatase activities capable of removing both the 3- and 5-phosphates from PtdIns(3,4,5)P_3 *in vitro* (2, 3, 7), though the kinetics of agonist-activated accumulation of PtdIns(3,4,5)P_3 and PtdIns(3,4)P_2 suggest that PtdIns(3,4)P_2 is the principal product of dephosphorylation of PtdIns(3,4,5)P_3 *in vivo* (2, 3). In contrast to animal cells plants do not appear to contain PtdIns(3,4,5)P_3 (8, 9). In plants PtdIns(3,4)P_2 is synthesized *in vivo* by 4- phosphorylation of PtdIns3P. Interestingly, a PtdIns3P 4-kinase activity, distinct from the PtdIns 4-kinase purified from erythrocytes, has been identified (10). The product of this activity occurs at very low levels in erythrocytes. It is currently not possible to define the contributions of PtdIns(3,4,5)P_3 5-phosphatase activities and PtdIns3P 4-kinase activities to PtdIns(3,4)P_2 production in unstimulated animal cells.

Plants, animals and yeast all contain phosphoinositide 3-kinase activities, enzyme activities capable of phosphorylating one or all of the lipids PtdIns, PtdIns4P and PtdIns(4,5)P_2. Among these, animals and yeast possess PtdIns-specific enzymes, PtdIns 3-kinases (11–13). It is likely that the plant enzyme (14, 15) is PtdIns-specific also. Yeasts do not appear to have PtdIns(3,4)P_2 or PtdIns(3,4,5)P_3. Thus it is possible that PtdIns(3,4)P_2 production and the substrate-cycling of this lipid detected in plants (16, 17) is evolutionarily unique and indicative of specific functions for PtdIns(3,4)P_2 in the plant kingdom.

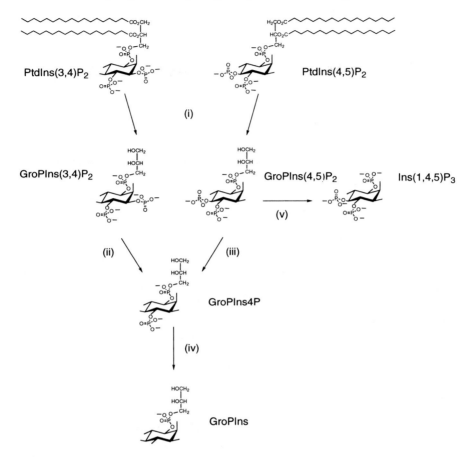

Figure 3. Routes of dissection of inositol phospholipids by chemical and enzymic means. (i) Deacylation, (ii) erythrocyte ghost inositol polyphosphate 3-phosphatase, (iii) erythrocyte ghost inositol·polyphosphate 5-phosphatase, (iv) alkaline phosphatase, (v) deglyceration. GroPIns, glycerophosphoinositol; GroPIns4P, glycerophosphoinositol 4-monophosphate; GroPIns (3,4)P_2, glycerophosphoinositol 3,4-bisphosphate; GroPIns (4,5)P_2, glycerophosphoinositol 4,5-bisphosphate; Ins(1,4,5)P_3, inositol 1,4,5-trisphosphate; PtdIns(3,4)P_2, phosphatidylinositol 3,4-bisphosphate; PtdIns(4,5)P_2, phosphatidylinositol 4,5-bisphosphate.

3. Experimental approaches to the study of turnover

A thorough description of inositol phospholipid turnover in stimulated cells requires knowledge of phospholipid turnover in the unstimulated state, that is, basal metabolism. Perhaps the most comprehensive description of basal inositol phospholipid metabolism has been provided by studies in human erythrocytes. Until very recently the erythrocyte was the only experimental

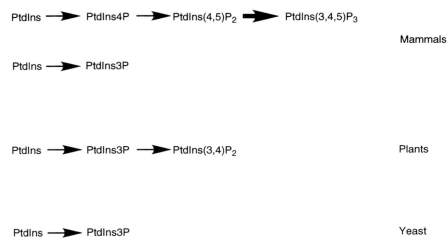

Figure 4. Kinase-mediated synthesis routes of 3-phosphorylated phosphatidylinositols in mammalian, plant and yeast cells, and for which there is strong metabolic evidence. The kinetics of agonist-driven accumulation of PtdIns(3,4)P_2 in mammalian cells suggests that this lipid is a product of dephosphorylation of PtdIns(3,4,5)P_3 by PtdIns(3,4,5)P_3 5-phosphatase (2, 3). The step denoted by the heavy arrow is acutely sensitive to activation in receptor protein tyrosine kinase and G-protein dependent manner (2, 5, 6). PtdIns, phosphatidylinositol (phosphomonoester phosphates identified relative to the D-I phosphodiester phosphate of PtdIns). PtdIns3P, phosphatidylinositol 3-monophosphate; PtdIns4P, phosphatidylinositol 4-monophosphate; PtdIns(3,4)P_2, phosphatidylinositol 3,4-bisphosphate; PtdIns(4,5)P_2, phosphatidylinositol 4,5-bisphosphate; PtdIns(3,4,5)P_3, phosphatidylinositol 3,4,5-trisphosphate.

system in which basal substrate cycling of inositol phospholipids had been demonstrated (18, 19). A distinction is made here between substrate cycling, which brings about equilibration of individual phosphate esters on the inositol ring (e.g. the 4- and 5-phosphates of PtdIns(4,5)P_2, i.e. between PtdIns4P and PtdIns(4,5)P_2), and basal turnover of the whole inositol phosphate headgroup, Ins(1,4,5)P_3, mediated by PtdIns(4,5)P_2-directed phosphoinositidase C (*Figure 5*). Furthermore, it should be made clear that substrate cycling is a metabolic definition born out of analysis of the labelling of individual phosphates *in vivo* and not an extrapolation of the kinase and phosphatase activities which have been identified *in vitro*. Thus, Müller and coworkers (18) and King and coworkers (19), cited in (20), were able to demonstrate substrate cycling of PtdIns4P and PtdIns(4,5)P_2 by analysis of the distribution of [^{32}P] orthophosphate label between the individual phosphates of PtdIns(4,5)P_2 during the approach to equilibrium.

In the context of cell stimulation, inositol phospholipid metabolism in mammalian erythrocytes, unlike that in their avian counterpart, is particularly inert. Also, it is not clear whether other agonist-responsive mammalian

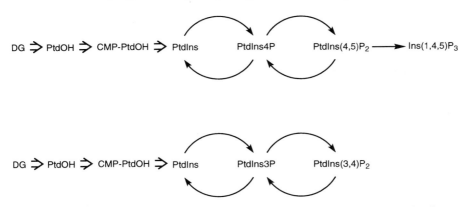

Figure 5. Routes of turnover of phosphomonoester phosphates and phosphodiester phosphates of inositol phospholipids in plant cells. Steps for which there is strong metabolic evidence and which in combination form potentially linked substrate cycles catalysed by inositol phospholipid kinase and monoesterase activities are indicated by curved arrows. Steps catalysed by phosphoinositidase C (substrate specificity unde-fined) for which only indirect or *in vitro* evidence exists are indicated by a single straight arrow. Steps considered to be responsible for the *de novo* synthesis of PtdIns and hence for the incorporation of the phosphodiester phosphate into PtdIns4P and PtdIns(4,5)P_2, and ultimately into the D-I phosphomonoester phosphate of Ins(I,4,5)P_3, are indicated by double arrows. The same steps are considered to be responsible for the incorporation of the phosphodiester phosphate into PtdIns3P and PtdIns(3,4)P_2. CMP-PtdOH, cytosine monophosphate phosphatidate; DG, diacylglycerol; Ins(I,4,5)P_3, inositol 1,4,5-trisphos-phate; PtdOH, phosphatidic acid; PtdIns, phosphatidylinositol (phosphomonoester phos-phates identified relative to the D-1 phosphodiester phosphate of PtdIns). PtdIns3P, phosphatidylinositol 3-monophosphate; PtdIns4P, phosphatidylinositol 4-monophos-phate; PtdIns(3,4)P_2, phosphatidylinositol 3,4-bisphosphate; PtdIns(4,5)P_2, phosphatidyl-inositol 4,5-bisphosphate.

cells show substrate cycling (20). Thus, the contribution of substrate cycling to activated inositol phospholipid turnover, conventionally focused on Ins(1,4,5)P_3 production, has until recently been undefined. Similarly, although basal substrate cycling has been described in the plant kingdom (16, 17), the lack of evidence for definable agonist-responsive changes in stereo-chemically identified Ins(1,4,5)P_3 production (other than by the use of Ins(1,4,5)P_3-receptor based binding assays on total cell extracts) does not allow an assessment to be made of the contribution of substrate cycling to stimulated Ins(1,4,5)P_3 production. Incidentally, the comparative simplicity of 3-phosphorylated lipid metabolism in the plant kingdom has facilitated the detailed description of basal substrate cycling of 3-phosphorylated lipids in plants (16, 17), something not yet possible for the animal kingdom.

Recent work (20) has shown that in agonist-stimulated cells, permeabilized neutrophils, PtdIns(4,5)P_2 production is matched to demand for inositol phosphate production principally by activation of PtdIns4P 5-kinase rather than by increased substrate cycling. Such a phenomenon is also evidenced in the data of Harden and coworkers (21), which indicates that in β-adrenergic receptor-stimulated turkey erythrocyte ghosts the linear accumulation of inositol phosphates could be maintained for periods up to 30 min in the presence of an ATP-regenerating system. Considering the foregoing, any analysis of inositol phospholipid turnover in plants must address or be mindful of the consequences of the metabolic processes described above for labelling of phospholipids.

3.1 Equilibrium labelling

Perhaps the most common approach to the study of inositol phospholipid turnover is equilibrium labelling of the inositol moiety of the headgroup. A precondition of such an approach is that cells are chronically labelled with a metabolic precursor, usually [³H]inositol, so that the pools of lipid are in equilibrium with the labelled metabolic precursor. Thus any changes in the levels of labelling of the inositol phospholipids (e.g. on cell stimulation) reflect changes in the mass of inositol phospholipid. It is important in such an approach that, though frequently not tested, the labelling of individual inositol phospholipids has reached a maximum before experimental manipulation of the system. The use of mutants deficient in inositol production and hence dependent on exogenous inositol is of great value in establishing equilibrium labelling (22).

We should be aware however that in equilibrium or non-equilibrium labelling situations stimulated turnover of inositol phospholipids may result in an increase, decrease or maintainence of the level of labelling of PtdIns(4,5)P_2, or even more complicated phenomena. Clearly, the most informative assessment of the consequences of turnover in this experimental approach is provided by quantification of the production of inositol phosphates. This is not of relevance to 3-phosphorylated lipids as these lipids are not substrates for phosphoinositidase C (3, 23, 24) and so strategies for studying the turnover of these lipids have to be concentrated on the lipid itself.

A slightly more complicated approach involves dual-labelling of the same chemical precursor with two isotopes. Thus Michell and coworkers (25) labelled cells to equilibrium with [¹⁴C]inositol and briefly labelled the cells with [³H]inositol prior to stimulation in an attempt to determine whether the inositol phospholipids recruited for inositol phosphate production on cell stimulation were typical of the bulk cellular population of inositol phospholipids or characteristic of a small hormone-sensitive subset.

3.2 Short-term (non-equilibrium) labelling strategies

Short term and non-equilibrium labelling strategies have been definitive in describing the routes of synthesis and turnover of individual inositol phospholipids in animal cells under stimulatory and non-stimulatory conditions. Thus in an early study King and coworkers (19) followed changes in

8

[^{32}P]orthophosphate labelling of the 4- and 5-monoester phosphates of human erythrocyte PtdIns(4,5)P_2 over a 7 h labelling period. At early times of labelling the 5- phosphate was most strongly labelled and by 7 h of labelling the 5-phosphate had come to equilibrium with that in the 4-position. Along with the work of Müller and coworkers (18) this study provided the first demonstration of basal substrate cyling of inositol phospholipids. Similar phenomena have been demonstrated in plants (16, 17). In addition, in the duckweed *Spirodela polyrhiza* L. an analysis of the labelling of the diester phosphate of PtdIns(4,5)P_2, largely independent of substrate-cycling associated changes in the monoester phosphates, illustrated basal turnover of the diester phosphate of PtdIns(4,5)P_2, presumably by phosphoinositidase C.

In an extrapolation of the strategy to very short-term labelling conditions and by analysis of the distribution of label among the individual phosphates of conventional and novel 3-phosphorylated lipids labelled from [^{32}P]orthophosphate, Stephens and coworkers (2) were able to determine the route of stimulated synthesis of PtdIns(3,4,5)P_3. It was subsequently shown that activated PtdIns(3,4,5)P_3 production is controlled in a G-protein dependent manner (5). The rationale for this experimental strategy has been described in detail (26). Briefly, providing that the specific activity of the γ-phosphate of the ATP pool is increasing during a labelling period, then any phosphate added at a particular point in a metabolic sequence will be more strongly labelled than any added before it and less strongly than any added after it. Thus, by estimation of the relative labelling of the different phosphate groups it is possible to determine the order of addition of the different phosphate groups to, in this case, the inositol moiety of the inositol phospholipid. This technique, allied to permeabilization of animal cells and labelling from [γ-^{32}P]ATP, has further confirmed that PtdIns(3,4,5)P_3 production, in both receptor tyrosine-kinase (27) and G-protein (28) dependent manner, is maintained in permeabilized cells.

4. Choice of experimental material

4.1 Whole plant/tissue

The choice of experimental material lies principally with the nature of the process which is under study—whether physiological, pathological, or developmental. Nevertheless, there are clear advantages for the researcher in choosing experimental systems which predominantly if not exclusively consist of a single cell type. Clearly, many developmental events in particular cells are influenced if not controlled by interaction with neighbouring cells of different type. From a developmental perspective it is frequently the case that the primordia of plant organs are a very small part of the plant organ and are difficult to isolate or are dysfunctional when isolated. In these cases some of the techniques described in this chapter, in particular the short-term non-equilibrium labelling strategies, confer the unique advantage that the experimental approach preferentially highlights the metabolism of the most active cells. Only the phospholipids, almost invariaby inositol phospholipids, of those cells showing the most rapid phospholipid turnover become labelled at short periods of labelling. Furthermore, conventional thinking has it that rapid basal turnover of molecules (29), and particularly substrate-cycling, is consistent with signalling roles or at least with the ability to respond to change in metabolic demand.

Ideally, the experimental material should be a homogeneous population of responsive cells. The population should be robust and easily manipulable and should demonstrate physiological or developmental responses which are easily quantified. Since plant cells appear to have low levels of PtdInsP_2s relative to other inositol phospholipids, when compared to many animal cells, the 'conventional' emphasis on PtdIns(4,5)P_2 and its contribution to signalling in animal cells means that large populations of plant cells will be required.

4.2 Suspension culture or protoplasts

Ease of manipulation dictates that the cells should preferably be in suspension culture and as monodisperse as possible. Alternatively, the use of protoplasts may appear at first to be ideal, though there is some concern as to whether treatment of plant cells with cell wall degrading enzymes during protoplast preparation activates inositol phospholipid metabolism (30). It is far from clear whether such events are general phenomena or are unique to the experimental system and the hydrolytic treatment used to obtain protoplasts. Nevertheless, many protoplasts show quantifiable responses to all manner of stimuli and some, such as those derived from stomatal guard cells and aleurone cells, have contributed new insights to our understanding of plant–hormone interaction. Suspensions of protoplasts have the advantage that the cells can be labelled uniformly, treatments can be administered to entire populations of cells simultaneously and labelling reactions can be stopped rapidly and synchronously. The cells are easily dispensed with air-displacement type pipettes.

The low level of labelling of PtdIns(4,5)P_2 relative to other phosphoinositides in plants (we do not know whether this is a reflection of chemical mass) has the consequence that studies of this lipid require prolonged labelling periods or large amounts of radiolabel. Where an investigation of the turnover of the monoester phosphates (as distinct from the whole headgroup) is required then only ^{32}P or ^{33}P labelling will suffice and for the former experimental protocols necessitate careful consideration regarding shielding from potentially harmful radiation. In these circumstances suspension cultures or protoplasts represent the best experimental material as they can be contained in thick walled screw-capped pyrex tubes which offer some shielding from radiation.

4.3 Use of permeabilized cells

One approach to minimizing the quantity of radioisotope used in labelling experiments is to label permeabilized cells directly with a more immediate metabolic precursor by permeabilization of the cell membrane and direct introduction of the precursor into the cytoplasm (27, 28). Perhaps the most obvious and practical precursor for inositol phospholipids is [γ-^{32}P]ATP, which is the immediate metabolic precursor of the monoester phosphates of all the inositol phospholipids. This precursor has proved to be rather less useful for labelling the diester phosphate but it labels the *de novo* precursor of PtdIns, PtdOH, very strongly. The corollary of permeabilization and the introduction of exogenous agents into cells is that some of the cellular contents are lost to the extracellular medium.

Cell permeabilization can be persistent or transient in nature and can be achieved in a number of ways (31). Persistent permeabilization can be achieved by treatment with plant glycosides such as digitonin (28, 31, 32), which forms complexes with unconjugated 3β-hydroxysterol components of membranes. Alternatively, bacterial toxins may be used: the α-toxin of *Staphylococcus aureus* (31) and streptolysin O (27, 28) intercalate into bilayers and form stable pores with a particular molecular cut-off. Hypotonic lysis of cells (31, 33, 34) creates single pores the size of which can be manipulated and maintained, particularly by the divalent cation composition of the medium. Glycol loading of cells and subsequent iso-osmotic lysis (31, 35) achieves the same results and is the basis of well tried methods for preparing erythrocyte ghosts. Transient permeabilization can be obtained by wash-out of some lytic agents or by e.g. dielectric breakdown (28, 31) of the cell membrane on exposure to a transient electric field (electroporation). Electroporation is commonly used for genetic transformation of cells.

Permeabilization is usually performed in an isotonic intracellular buffer designed to mimic the cytoplasmic concentration of important cellular components. Permeabilization has also proved to be fully compatible with short-term non-equilibrium labelling strategies and under appropriate controls of physiological function has been used in investigations of receptor agonist-induced changes in inositol phospholipid metabolism of various animal cells (27, 28).

The presence of a rigid cell wall around plant cells complicates some permeabilization strategies. Permeabilization of plant cells is usually performed after enzymic removal of the cell wall. The protoplasts released must be maintained within a narrow range of osmolarity if osmotic excursions which are normally constrained in the intact cell by the cell wall are not to become lytic.

4.3.1 Transient permeabilization of plant protoplasts by electroporation

A method for preparing protoplasts from leaf mesophyll cells is detailed in *Protocol 1*.

Protocol 1. Preparation of mesophyll protoplasts of *Commelina communis* L.

Equipment and reagents

- Metal test sieves, 100 μm and 50 μm (Endecotts Ltd.)
- Orbital shaker, Variomag Model R100 (Luckhams Ltd.)
- Cellulysin (Calbiochem)
- Holding medium: 400 mM mannitol, 1 mM KCl, 0.1 mM CaCl₂, 0.5 mM ascorbic acid, 0.5 mM dithiothreitol, 10 mM Mes buffer adjusted to pH 5.5 with KOH
- Cellulase 'Onozuka' RS (Yakult Pharmaceutical Ind. Company)
- Pectolyase Y 23 (Sigma)
- Digestion medium: 2% w/v cellulase, 2% w/v cellulysin, 0.026% w/v pectolyase Y 23, 0.26% w/v bovine serum albumin, 10 000 U penicillin and 10 000 U streptomycin per 30 ml, prepared in holding medium

Method

1. Detach the youngest fully expanded leaves of 5–6 week old plants of *Commelina communis* L. and remove the mid rib with a razor blade.

2. Turn the leaf so that the abaxial surface is upward and bend one end of the leaf section up and over the body of the leaf.

3. Using a pair of fine forceps at the bend in the leaf section scrape with the forceps and expose the epidermis.

4. Peel the epidermis longtitudinally down the leaf and discard.

5. Lay the leaf lamina abaxial surface down on holding medium contained in a 9 cm diameter polystyrene Petri dish.

6. Continue until the surface of the solution is covered.

7. Replace the holding medium with 4 ml of freshly prepared digestion medium and incubate in the dark at 30 °C for 2 h with gentle agitation.

8. Following incubation 'tap' the Petri dish gently on the bench to release the protoplasts.

9. Filter the cell suspension through the 100 μm and 50 μm test sieves.

10. Repeat step 9 three times.

11. Centrifuge the filtered cell suspension at 100 × *g* for 5 min.

12. Wash the protoplast pellet by resuspending the protoplasts in 5 ml of holding medium and repeat step 11.

13. Repeat step 12.

14. Resuspend the protoplasts in electroporation medium (*Protocol 2*).

A permeabilization regime which has been used to label all the inositol phospholipids commonly found in plant cells and which is effective with protoplasts from moss, aleurone cells, stomatal guard cells and mesophyll cells is described in *Protocol 2*.

Protocol 2. Electroporation of mesophyll protoplasts

Equipment and reagents

- Electroporation apparatus (Pro-Genetor II, PG220P-2.5 electrode, Hoefer)
- Electroporation medium: 450 mM mannitol, 20 mM KCl, 10 mM MgCl$_2$, 60 μM MgATP, 0.3 mM CaCl$_2$, 1 mM EGTA, 20 mM Hepes, pH 7.6 at 20°C
- 4 Well Multidish (Nunc)
- [γ-^{32}P]ATP, specific activity 3000 Ci mmol^{-1} (Amersham)

Method

1. Dispense 0.3 ml of a protoplast suspension (3 $\times$ 10^6 cells/ml) in electroporation medium into the well of a 4 Well Multidish.

2. Add 30 μCi of ATP.

3. Electroporate the protoplasts with two 100 μF discharges of 5 msec duration at a field strength of 500 V cm^{-1} and 15 sec interval between pulses.

4. Incubate the cells for 5 min.

5. Add 0.6 ml of ice-cold 5% HClO$_4$ and transfer the Multidish to ice for 15 min.

6. Transfer the cell extract to a pre-cooled microcentrifuge tube and centrifuge at 7000 $\times$ *g* for 5 min at 4°C in a refrigerated microcentrifuge.

7. Remove the supernatant and discard.

8. Extract lipids from the cell debris (*Protocol 4*).

---

5. Extraction of inositol phospholipids from plant tissues

The special problem of extracting acidic phospholipids from plant tissues is one which has received considerable attention. The adoption of acidified solvents and the inclusion of chelators of divalent cations has reduced losses of inositol phospholipids. What has received rather less attention is the possibility of activation of phospholipases, particularly phospholipase D, by organic solvents in extraction regimes in which cells or tissues are extracted directly in solvent without steps to inactivate enzyme activity first. Only very recently has a signalling role for phospholipase D been identified in plant tissues. This was evidenced primarily by the activation of the well documented transphosphatidylase activity of phospholipase D by primary alcohols in pre-labelled tissue (36). The fact that plants contain phospholipases which can compromize recoveries of phospholipids has been known for considerably longer (37, 38).

Whatever the method of extraction, it is essential that the plant, tissue or cell suspension is in as fragmented a state as possible before extraction. The presence of a cell wall, though permeable to solvent and therefore probably not a barrier to extraction of lipids from monodisperse cell suspension, requires that cultures or tissues with considerably more 'structure' are homogenized first to ensure rapid enzyme inactivation. Perhaps the best method for inactivating enzyme activity during homogenization is to cool the tissue with liquid nitrogen and then to grind the tissue to a fine powder in a cooled mortar and pestle.

Our preferred method of homogenization (*Protocol 3*) for tissue and suspension cultures is to extract the ground tissue or filtered cell suspension with perchloric acid containing an aliquot of a hydrolysate of phytate. One of the advantages of such a regime is that this is the preferred method for extraction of inositol phosphates, the hydrolysate minimizes losses due to non-specific binding (39) while the phospholipids can be efficiently extracted from the acid-insoluble cell debris.

Extraction of inositol phospholipids from protoplasts can be initiated by the addition of ice-cold perchloric acid and rapid mixing followed by extraction of the lipids from the perchloric acid-insoluble cell debris (*Protocol 4*). This, in our hands, gives a much more consistent recovery of phospholipids than extraction directly into organic solvent.

Protocol 3. Homogenization of plant tissues

Equipment and reagents
- Mortar and pestle
- Gel tips, 'long-reach' pipette tips (Alpha Laboratories Ltd.)
- Perchloric acid (5% w/v)
- Liquid nitrogen
- Phytate hydrolysate[a]

Method

1. Blot dry the labelled tissue on filter paper[b].

2. Cool the tissue in liquid nitrogen.

3. Transfer the tissue to a precooled mortar and grind the tissue to a fine powder with a precooled pestle.

4. Add 1 ml of 5% perchloric acid and add 50 μl of a phytate hydrolysate containing 50 μg of phytate-derived phosphorus[c].

5. Allow the pestle to warm slightly and grind the frozen solution of perchloric acid so that the perchloric acid is dispersed around the entire mortar.

6. Allow the perchloric to thaw and leave the tissue to extract in the perchloric acid for 15 min with occasional grinding[d].

7. Transfer the perchloric acid extract including the insoluble material to

---

a precooled 2 ml microcentifuge tube and centrifuge in a refrigerated microcentrifuge at 7000 × g for 5 min.

8. Remove the supernatant with a Gilson-type air displacement pipette[e] and retain for analysis of inositol phosphates or discard.

9. Extract the phospholipids from the cell debris (*Protocol 4*).

[a]Prepare the hydrolysate of phytate according to the method of Wreggett and Irvine (39).
[b]Suspension cultures of cells can be dried by vacuum filtration on a Buchner funnel and flask connected to a water pump.
[c]The perchloric acid and the solution of phytate hydrolysate will freeze on contact with the mortar.
[d]The mortar and contents should not be allowed to warm above 4°C so it is necessary to transfer the mortar to ice.
[e]It is possible to remove the last few μl of supernatant without disturbing the pelleted cell debris using 'long-reach' pipette tips.

Inositol phospholipids can be conveniently extracted from the pelleted cell debris according to *Protocol 4*.

Protocol 4. Extraction of inositol phospholipids from plant tissues[a]

Equipment and reagents

- 1 ml glass Tuberculin syringes (Richardsons of Leicester Ltd.)
- Extraction and partitioning solvents:
  (i) Chloroform: methanol, 1:2, v/v
  (ii) 2.4 M HCl
  (iii) 1 mM EDTA disodium salt
  (iv) Chloroform
- 3 inch-long luer hub Hamilton HPLC needle (Phase Separation Ltd.)
- Screw-capped Pyrex culture tubes, 16 mm o.d. × 100 mm (Bibby Sterilin Ltd.)
- Organic phase wash: methanol: 1 M HCl, 1:1, v/v, 1 mM CDTA, 1 mM *myo*-inositol, 1 mM orthophosphoric acid

Method

1. Resuspend the pelleted cell debris (*Protocol 3*) in 1.5 ml of chloroform: methanol and transfer to a pyrex tube.

2. Cap the tube and mix vigorously on a vortex-mixer.

3. Sonicate the tube and contents in a bath sonicator for 5 min.

4. Wash out the microcentrifuge tube from which the cell debris has been removed with 0.5 ml of HCl and transfer the washings to the pyrex tube.

5. Repeat step 3.

6. Repeat in sequence steps 4 and 3, substituting 0.5 ml of 1 mM EDTA for HCl.

7. Repeat in sequence steps 4 and 3, substituting 0.5 ml of chloroform for HCl in step 4.

Protocol 4. *Continued*

8. Centrifuge the tube and contents at 1000 × *g* for 3 min.

9. Remove the lower organic phase with the glass syringe and HPLC needle to a second pyrex tube.

10. Add 0.5 ml of chloroform and repeat in sequence steps 2, 3, 8, and 9.

11. Repeat step 10.

12. Add 2 ml of organic phase wash solution to the pooled organic phase.

13. Mix the contents, centrifuge at 1000 × *g* for 3 min and remove the upper aqueous phase to waste.

14. Repeat step 13 but leave some of the upper aqueous phase behind in the tube.

15. Using a clean syringe and needle remove the organic phase, from under the remaining aqueous phase, to a third pyrex tube.

16. Reduce the lipid to dryness under a gentle stream of nitrogen.

[a]Method modified from Boss (40) and Stephens *et al.* (2).

6. Identification of inositol phospholipids

Perhaps the most serious and obvious obstacle to the study of inositol phospholipid turnover in plants is the authentication of the identity of such components. The existence of isomeric forms of PtdIns$P$ and PtdInsP_2 does not allow for simple classifications of phosphoinositides into PtdIns$P$ and PtdInsP_2 species. It is not safe to assume that the 3-phosphorylated lipids, PtdIns3P and PtdIns(3,4)P_2, are minor components in comparison to PtdIns4P and PtdIns(4,5)P_2, nor is it safe to assume that these are the only inositol phospholipids in plant tissues. The problem is compounded by the existence in plants of other lipids which not only incorporate [³H]inositol and/or [³²P]orthophosphate, and often rapidly, but also co-chromatograph with authenticated inositol phospholipids on the TLC systems commonly used to separate lipid extracts (1, 41). These lipids may belong to a class of lipids called sphingolipids which are relatively poorly characterized in plants and for which the nomenclature is particularly confused. Moreover the considerable influence in TLC of acyl substituents can have the consequence that inositol phospholipid standards obtained from one species or one kingdom do not run with inositol phospholipids from another. Thus, given that co-migration of 'unknowns' with authentic standards is the raison d'être of identification it is also worth considering that it is very easy to arrive at suboptimal chromatographic conditions. Consequently, the co-migration of an 'unknown' with an authentic standard is only of value if other likely 'candidates' can be excluded on chromatographic grounds. Similarly, it is far safer to include standards labelled with another isotope with the 'unknown' in the same chromatographic run than it is to assume that the separation parameters are identical in separate runs of standard and 'unknown'. Little is known of the manner of turnover or remodelling of fatty acid moieties of inositol phospholipids. In any case, the general rationale for analytical strategies is based around the headgroup, which also reflects the fact that it is the main functional constituent.

6.1 HPLC analysis of phospholipid headgroups

For plant studies the combination of [^{3}H]inositol labelling and anion exchange HPLC of the water-soluble products of deacylation offers unparalled resolution of the different inositol glycerophospholipids. Thus in a single HPLC run all of the inositol labelled components of the 'classical' PtdIns cycle, that is, PtdIns, PtdIns4P and PtdIns(4,5)P_2, can be distinguished from the 3-phosphorylated lipids, PtdIns3P, PtdIns(3,4)P_2 and PtdIns(3,4,5)P_3, of which only the latter has not yet been detected in plants. Where plant cells are labelled from [^{32}P]orthophosphate the situation is compromized by the co-elution of GroPGro, derived from phosphatidylglycerol, a major component of thylakoid membrane, and GroPIns derived from PtdIns (17). Generally, the emphasis placed on the substituted phosphatidylinositols renders the problem of little importance. The application of ^{32}P labelling does have added benefits in that PtdOH, another component of the PtdIns cycle, is identified as glycerol phosphate (GroP), and so is diacylglycerol pyrophosphate, identified as glycerol pyrophosphate (GroPP). The latter lipid, described *in vitro* (42, 43), has been identified in permeabilized plant cells. This lipid has not been found in animal cells. In the context of the present chapter both GroP and GroPP are resolved by anion exchange HPLC from the products of deacylation of all of the inositol phospholipids mentioned above (*Figure 6*).

6.1.1 Deacylation of phospholipids

Removal of the acyl chains from glycerophospholipids can be achieved in a number of ways. O–N transacylation with monomethylamine (44) has largely replaced alkaline methanolic hydrolysis, conferring the advantage of diminished side reactions. The technique can be applied to total lipid extracts, and the water-soluble products of interest readily separated from the fatty amides by simple two-phase partition (*Protocol 5*). The headgroups are resolved on anion exchange HPLC columns (*Protocol 6*).

6.1.2 HPLC of glycerophosphoinositol phosphates

The water-soluble products of lipid deacylation are resolved on anion exchange HPLC (*Protocol 6*). A separation of the products of deacylation of a [^{3}H]inositol-labelled lipid extract from *S. polyrhiza* is shown in *Figure 6*.

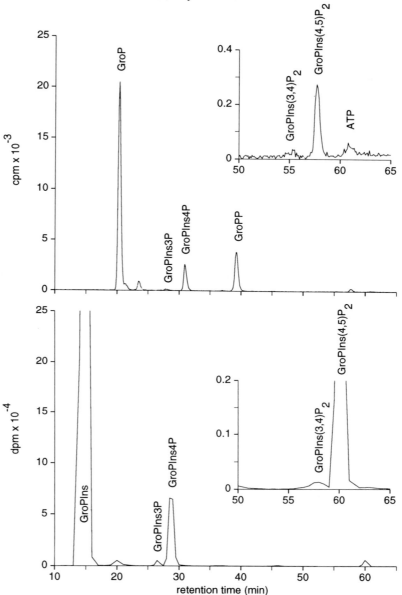

Figure 6. Partisphere SAX HPLC separation of the water-soluble products of deacylation of lipids from [γ-^{32}P]ATP-labelled permeabilized mesophyll protoplasts of *C. communis* (upper panel) and [^{3}H]inositol-labelled *S. polyrhiza* plants (lower panel). GroP, glycerol phosphate; GroPIns, glycerophosphoinositol; GroPP, glyceropyrophosphate; GroPIns3P, glycerophosphoinositol 3-monophosphate; GroPIns4P, glycerophosphoinositol 4-monophosphate; GroPIns(3,4)P_2, glycerophosphoinositol 3,4-bisphosphate; GroPIns(4,5)P_2, glycerophosphoinositol 4,5-bisphosphate.

Protocol 5. Deacylation of phospholipids[a]

Equipment and Reagents

- Screw-capped Pyrex culture tubes, 16 mm o.d. × 100 mm (Bibby Sterilin Ltd.)
- Deacylation reagent: monomethylamine, MeOH, water, BuOH, 5:4:3:1, v/v

- Monomethylamine, 40% aq. soln (Aldrich)
- Wash solution: BuOH, Petroleum Ether 40–60 °C fraction, ethyl formate, 20:4:1, v/v

Method

1. Transfer the washed lipid extract to a pyrex tube and reduce to dryness under a stream of nitrogen[b].

2. Add 1 ml of Deacylation reagent per 2 × 10^6 cells, mix vigorously and sonicate in a bath sonicator for 2 min to aid dissolution of the dried lipid film.

3. Incubate at 53 °C for 45 min.

4. Add 1.2 ml of water and allow to cool to room temperature. Add 1.2 ml of Wash reagent, mix vigorously and leave to stand for 5 min before centrifuging at approx. 1000 × *g* for 3 min.

5. Remove the upper phase with a pasteur pipette or a Gilson-type pipette and discard.

6. Add 1.2 ml of Wash reagent to the aqueous phase, mix vigorously and leave to stand before centrifuging (step 4).

7. Remove the lower aqueous phase with a Hamilton-type glass syringe or a 1 ml Tuberculin syringe fitted with an HPLC needle to a second tube for freeze-drying[c].

8. Snap-freeze the tube and contents in liquid nitrogen and freeze-dry the sample.

[a] Method of Hawkins *et al.* (45).
[b] The sample tube can be warmed gently eg. at 40 °C on a heating block to aid drying of the sample.
[c] By leaving a volume of the upper phase above the lower aqueous phase it is easier to identify the phase boundary.

Protocol 6. HPLC of glycerophosphoinositols[a]

Equipment and reagents

- Binary HPLC system fitted with 1 ml injection loop[b]
- UV detector set at 254 nm
- Optiphase HiSafe III scintillation fluid (Wallac UK)
- Partisphere SAX HPLC column, 25 cm × 4.6 mm (Whatman)[c]

- Fraction collector
- Buffer reservoirs containing A, water; B, 1.25 M $(NH_4)_2HPO_4$ adjusted to pH 3.8 with H_3PO_4 at 20 °C[d]
- Nucleotide markers[e]; 1 mM AMP, 1 mM ADP and 1 mM ATP

19

Protocol 6. *Continued*

Method

1. Take up the freeze-dried deacylate in 0.9 ml of water. Mix vigorously and sonicate briefly (*Protocol 5*).

2. Add 10 μl of each of the nucleotide markers.

3. Transfer the contents of the tube to a microcentrifuge tube and centrifuge the sample at full speed for 5 min in a microcentrifuge.

4. Remove the supernatant to a dust-free tube.

5. Program the HPLC machine to deliver the following gradient at a flow rate of 1 ml min^{-1}: 0 min, 0% B; 5 min, 0% B; 45 min, 12% B; 52 min, 20% B, 64 min, 100% B; 75 min, 100% B.

6. Pre-equilibrate the column with water at 1 ml min^{-1} for 75 min prior to injection of the sample.

7. Flush the injection port and loop with water, load the sample into the injection loop and inject the sample[f].

8. Monitor the absorbance at 254 nm and collect 0.5 min fractions.

9. Add 1 ml of water and 5 ml of Optiphase Hisafe III scintillation fluid to the samples.

10. Measure radioactivity by scintillation counting[g].

[a] Method of Stephens *et al.* (46)
[b] The minimal configurational requirement of the HPLC sytem for the simple gradient elution conditions of this separation is a binary pump.
[c] The use of disposable guard cartridges (Whatman AX cartridge) is recommended to prolong the life of the column.
[d] The buffers should be degassed either with helium during HPLC or under reduced pressure immediately prior to HPLC.
[e] The inclusion of nucleotide markers facilitates monitoring of the HPLC run.
[f] Ensure that the injection loop overflow is above the height of the injection port and collect overflow from injection for safe disposal.
[g] Alternatively, count a small aliquot of each fraction collected. ^{32}P labelled fractions can be counted directly without scintillant by Cerenkov counting.

6.1.3 Desalting of HPLC fractions

Further confirmation of the identity of the inositol phospholipids present in lipid extracts can be obtained by deglyceration of glycerophosphoinositols obtained above (section 6.1.2). Prior to deglyceration it is necessary to desalt the HPLC fractions. This is achieved on Dowex AG 1X8 formate form resin. The resin can be bought in the formate form or alternatively can be converted to the formate form by washing with 2 M ammonium formate. Before use the resin should be washed extensively with water (*Protocol 7*).

Protocol 7. Desalting of HPLC-purified glycerophosphoinositols[a]

Equipment and reagents
- Dowex AG 1X8 formate form resin (Bio Rad)
- Polypropylene mini-columns (Bio Rad)

Method

1. Wash 5 g of resin in 100 ml of 2 M ammonium formate for 30 min.
2. Allow the resin to settle and decant off the supernatant.
3. Dispense aliquots (approx. 0.4 ml) of the slurry into polypropylene mini-columns (10 ml volume) and wash the resin (0.2 ml settled volume) with 20 ml of water[b].
4. Pool the relevant HPLC fractions, dilute 10–20 fold with water and adjust to pH 6–7 with triethylamine.
5. Apply the sample to the column[c].
6. Elute with 10 ml of 100 mM ammonium formate, 50 mM formic acid[d]. This washes off P_i.
7. Cap the column and resuspend the resin with 1 ml of 1 M ammonium formate, 0.1 M formic acid. Leave to stand for 20 min and drain into a 20 ml polypropylene scintillation vial.
8. Repeat step 7 and pool the two eluates[e].
9. Snap-freeze the scintillation vial and contents in liquid nitrogen and freeze-dry until the white residue of ammonium formate has disappeared[f].

[a] Method modified from Stephens *et al.* (50).
[b] The columns can be prepared in bulk and stored for several months at 4°C.
[c] Apply the sample to the capped column and when the column bed has settled remove the cap and allow the column to drain.
[d] Allow the solution to drain straight through the column.
[e] The remaining traces of the sample can be recovered from the column by centrifuging the column and the 20 ml scintillation vial in a bench centrifuge at 100 × *g* for 3 min.
[f] The whole freeze-drying process can be speeded up by snap-freezing the sample on the surface of a rotating round-bottomed pyrex flask before the freeze-drying step.

6.1.4 Deglyceration of glycerophosphoinositols

Further confirmation of the identity of inositol phospholipids can be obtained by removal of the glycerol moiety with sodium periodate, described initially by Brown and Stewart (47) (*Protocol 8*) and subsequent chromatography on a Partisphere SAX HPLC column. The removal of the glycerol moiety converts the phosphodiester to a monoester and the subsequent increase in charge and polarity of the molecule is reflected in an increased retention time on SAX HPLC (*Protocol 6*).

Protocol 8. Deglyceration of glycerophosphoinositols[a]

Equipment and reagents
- Dowex AG 50W resin (Bio Rad)
- Polypropylene mini-columns (Bio Rad)
- 10 mM sodium *m*-periodate
- 500 mM ethanediol
- 0.5% dimethylhydrazine adjusted to pH 4.5 with formic acid[b]

Method

1. Take up the freeze-dried and desalted glycerophosphoinositol in 0.5 ml of 10 mM sodium periodate and incubate at 25°C for 30 min.

2. Add 0.5 ml of 500 mM ethanediol to quench unreacted periodate and incubate at 25°C for 30 min.

3. Add 1.8 ml of 0.5% dimethylhydrazine to eliminate aldehydes and incubate at 25°C for 3 h.

4. Apply the reaction products to a column (3 ml bed volume) of Dowex AG 50W resin, H$^+$ form[c,d], and collect the eluate.

5. Wash the column with 6 ml of water, pool the washings with the eluate from step 4. and freeze-dry the pooled sample (*Protocol 7*).

[a]Method from Whitman *et al.* (48).
[b]Dispense dimethylhydrazine in a fume hood.
[c]The resin should be prewashed in sequence with 2 M HCl and then several times with water.
[d]The column is very slow to drain. An alternative procedure is to mix the sample with resin for 30 min and then to centrifuge the sample briefly to separate the resin from the supernatant.

6.1.5 HPLC of the products of deglyceration of glycerophosphoinositols

The products of deglyceration of glycerophosphoinositols are inositol phosphates (*Figure 3*). Thus, GroPIns(4,5)P_2 is converted to Ins(1,4,5)P_3 and GroPIns(3,4)P_2 to Ins(1,3,4)P_3. Ins(1,3,4)P_3 and Ins(1,4,5)P_3 can be separated on Partisphere SAX HPLC columns by isocratic elution with phosphate buffers (*Protocol 9*). HPLC analysis of the inositol phosphate products of deglyceration of glycerophosphoinositols confers an advantage over analysis of the glycerophosphoinositols in that authentic standards of [^{3}H]Ins(1,4)P_2, Ins(1,3,4)P_3 and Ins(1,4,5)P_3 are commercially available, e.g. from NEN-Dupont or Amersham International. The inositol bisphosphate products of deglyceration of GroPIns3P and GroPIns4P, Ins(1,3)P_2 and Ins(1,4)P_2, can also be resolved isocratically (*Protocol 9*) and [^{3}H]Ins(1,3)P_2 can be prepared from [^{3}H]Ins(1,3,4)P_3 by alkaline hydrolysis (*Protocol 10*).

Protocol 9. Isocratic separation of inositol bis-and trisphosphates on Partisphere SAX HPLC columns[a]

Equipment and reagents

- HPLC system, Partisphere SAX HPLC column, UV detector and fraction collector (*Protocol 6*)
- Nucleotide markers (*Protocol 6*)

- Buffer reservoirs containing A, water; B, 240 mM NaH_2PO_4 (bisphosphates), 550 mM NaH_2PO_4 (trisphosphates)

Method

1. Take up the desalted sample in 0.9 ml of water, add nucleotide markers and centrifuge (*Protocol 6*).

2. Inject the sample onto a column pre-equilibrated with phosphate buffer (240 mM NaH_2PO_4, bisphosphates; 550 mM NaH_2PO_4, trisphosphates).

3. Collect fractions and measure radioactivity as described in *Protocol 6*.

[a] Method of Wreggett and Irvine (49).

Isocratic HPLC on Partisphere SAX columns can separate all of the isomeric species of inositol bis- and trisphosphates which are obtained on deacylation and deglyceration of the PtdIns$P$ and PtdInsP_2 species identified to date in plants.

6.1.6 Alkaline hydrolysis of inositol bis- and trisphosphates

The availability of commercial preparations of various inositol bis- and trisphosphates affords simple chemical means of preparing derivatives of these. Thus, alkaline hydrolysis of [^{3}H]Ins(1,3,4)P_3 yields a mixture of monophosphates and [^{3}H]Ins(1,3)P_2, [^{3}H]Ins(1,4)P_2 and [^{3}H]Ins(3,4)P_2. These three bisphosphates are all resolved under the isocratic separation conditions detailed in *Protocol 9*. Alkaline hydrolysis can be performed in flame sealed glass tubes with conc. ammonia at 110°C (50). The potentially explosive nature of such treatments requires special safety precautions to be taken. Alternatively, hydrolysis can be achieved in aqueous solution in a bench top autoclave (*Protocol 10*). Alkaline hydrolysis generates a mixture of inositol, inositol monophosphates and inositol bisphosphates and unreacted trisphosphate. The three bisphosphates, Ins(1,3)P_3, Ins(1,4)P_3 and Ins(3,4)P_3 are obtained in approximately equal amounts and the overall yield of bisphosphates is approximately 30%.

Protocol 10. Alkaline hydrolysis of inositol phosphates[a]

Equipment and reagents
- Bench top autoclave (Macarthy's Surgical Ltd.)
- Buffer, 100 mM Na_2HPO_4 adjusted to pH 10.5 with NaOH

Method
1. Take up the [3]H- or ^{32}P-labelled inositol phosphate in 0.2 ml of buffer.
2. Transfer to a pyrex test tube and cap the tube with an autoclave compatible lid.
3. Autoclave the tube and contents at 120 °C for 3 h.
4. Desalt the sample on Dowex AG 1X8 resin (*Protocol 7*)[b].

[a] Method of Stephens (51).
[b] Alternatively, a small amount of the autoclaved sample can be used to 'spike' the unknown sample before HPLC analysis.

---

7. Dissection of glycerophosphoinositol phosphates

The techniques described in this section are designed to determine the distribution of label among individual phosphate groups of inositol phospholipids labelled either directly from ^{32}P]ATP in permeabilized cells or from ^{32}P]orthophosphate in intact cells and tissues. Dissection of inositol phospholipids can be undertaken at the level of the water-soluble deacylation products or subsequently at the level of the deglycerated products (2, 5, 8, 16, 17, 18, 46, 52).

7.1 Glycerophosphoinositol monophosphates (GroPInsPs)

Dissection of glycerophosphoinositol monophosphates (GroPInsPs) is achieved with alkaline phosphatase (*Protocol 11*). Commercial preparations of alkaline phosphatase are not necessarily entirely specific for monoester phosphates. The one used in *Protocol 11* also attacks diester phosphates but with a much lower activity. Thus, if the reaction is allowed to proceed too far both the diester and the monoester phosphates of GroPInsP are lost as inorganic phosphate (P_i). By the inclusion in the assay of a 'metabolic trap' of unlabelled glycerophosphoinositol (GroPIns) it is possible to limit attack on labelled GroPInsP to the monoester thus yielding GroPIns and P_i as products from which the labelling of the diester and monoester phosphates respectively is obtained. The products are separated on Partisphere SAX HPLC (*Protocol 6*). The inclusion at the point of dissection of an internal standard of HPLC-purified ^{3}H]GroPInsP affords an added safeguard in that it allows the experimenter to verify that the reaction has not gone too far. Thus the

24

presence of [³H]inositol, which elutes at the solvent front, indicates attack on the diester. Hence in this case a simple comparison of the ³²P:³H ratio of the GroPIns*P* starting material and the GroPIns product yields the relative labelling of the diester and monoester phosphates.

Protocol 11. Dissection of GroPIns*P* species with alkaline phosphatase[a]

Equipment and reagents

- Thin-stemmed long-reach pH electrode (Type CMAWL S7/3.7/180, Russell pH Ltd.)
- Buffer: 20 mM Hepes adjusted to pH 7.5 with NaOH, 1 mm MgCl₂

- Alkaline phosphatase (Sigma P 5521)
- 10 mM glycerophosphoinositol (Sigma)
- 70% v/v perchloric acid

Method

1. Take up the desalted [³²P]GroPIns*P* and [³H]GroPIns*P* in 0.5 ml of buffer and transfer the sample to a microcentrifuge tube[b].

2. Adjust the sample to pH 7.5 with KOH added a few μl at a time[c].

3. Add GroPIns to a final concentration of 0.1 mM.

4. Add 2 Units of alkaline phosphatase (Units defined by Sigma) and incubate at 25°C for 10 min[c].

5. Stop the reaction by the addition of 10 μl of 70% perchloric acid.

6. Incubate the sample on ice for 15 min and centrifuge at full speed in a microcentrifuge for 3 min.

7. Remove the supernatant, add nucleotide markers, and apply the sample to a Partisphere SAX HPLC column (*Protocol 6*).

[a]Method modified from Stephens *et al.* (46).
[b]It is best to co-purify the internal standard with the [³²P]GroPIns*P* sample to ensure that the ³²P:³H ratio of the GroPIns product is not compromised by breakdown of the internal [³H]GroPIns*P* standard to [³H]GroPIns on storage. Alternatively, the purity of the standard can be checked by HPLC prior to dissection of the [³²P]GroPIns*P* sample.
[c]The presence of traces of ammonium formate left-over from the desalting step requires that the pH is adjusted to 7.5.
[d]The dilution of alkaline phosphatase required is best determined in a trial run with [³H]GroPIns4P which can be prepared by deacylation of commercially available [³H]PtdIns4P (*Protocol 5*). Further 10-fold dilutions allow the determination of a working range of enzyme concentration.

7.2 Glycerophosphoinositol bisphosphates (GroPIns*P₂*s)

Dissection of glycerophosphoinositol bisphosphates can be achieved by the use of enzyme activities which are specific for monoester phosphate groups in particular positions on the inositol ring or with activities which show little discrimination between monoester phosphates but which discriminate between the monoester and diester phosphates. The inositol polyphosphate 5-phosphatase and 3-phosphatase activities of erythrocytes (53) have been used frequently to determine the distribution of label among the individual monoester phosphates of inositol phospholipids (2, 5, 16, 17, 18, 46, 52). The techniques described in this section afford a powerful strategy for studying the turnover of individual phosphate monoesters and the diester of inositol phospholipids and should allow a full assessment and description of inositol phospholipid turnover in plants to be made under a variety of 'stimulatory' regimes.

7.2.1 Dissection of GroPIns(4,5)P_2 with alkaline phosphatase

Alkaline phosphatase preferentially attacks phosphate monoesters that are without flanking phosphates on the neighbouring carbon atoms. Thus alkaline phosphatase can be used to generate Ins(4,5)P_2 from Ins(1,4,5)P_3. When the phosphate in the 1-position of the inositol ring is a diester as in GroPIns(4,5)P_2 the monoester phosphates in positions 4- and 5- are preferentially attacked. Thus, by the inclusion of a metabolic trap of unlabelled GroPIns as in *Protocol 11*, it is possible to determine the labelling of the diester phosphate in GroPIns(4,5)P_2 and hence in PtdIns(4,5)P_2. Label in the diester phosphate is added by diacylglycerol kinase in the reaction which generates the PtdOH precursor of PtdIns. Consequently, changes in the labelling of the diester of PtdIns(4,5)P_2 represent flux in the sequence DG > PtdOH > CMP-PtdOH > PtdIns > PtdIns4P > PtdIns(4,5)P_2, independent of any rearrangement of the monoester phosphates by kinase–monoesterase substrate cycles.

Protocol 12. Estimation of labelling of the diester phosphate of PtdIns(4,5)P_2[a]

Equipment and reagents
- Buffer (*Protocol 11*)
- GroPIns (*Protocol 11*)
- Alkaline phosphatase (*Protocol 11*)

Method

1. Take up the desalted [^{32}P]GroPIns(4,5)P_2 and [^{3}H]GroPIns(4,5)P_2[b] in 0.5 ml of buffer and transfer the sample to a microcentrifuge tube.

2. Adjust the sample to pH 7.5, add GroPIns and alkaline phosphatase and incubate the sample at 25°C for 10 min (*Protocol 11*).

3. Stop the reaction and process the sample for HPLC (*Protocol 11*).

4. Perform HPLC according to *Protocol 6*.

[b]Method modified from Stephens *et al.* (46).
[b][^{3}H]GroPIns(4,5)P_2 is prepared by deacylation of [^{3}H]PtdIns(4,5)P_2 which can be obtained commercially (Amersham).

7.2.2 Dissection of glycerophosphoinositol bisphosphates with erythrocyte ghost inositol polyphosphate phosphatase activities

Erythrocyte ghosts can easily be prepared from human erythrocytes. They provide a source of inositol polyphosphate phosphatase activities which can be used to remove phosphate monoester groups selectively from the 3- and 5-positions of the inositol ring of inositol phospholipids. The considerable power of resolution of stereoisomers afforded by Partisphere SAX HPLC ensures that all of the potential products of dissection of individual inositol phospholipids can be resolved in a single HPLC run. Thus by the inclusion of ^{3}H standards in the HPLC step the products of dissection can be unequivocally identified and the ^{32}P-labelling of individual phosphate groups determined.

(i) Preparation of erythrocyte ghosts

Erythrocyte ghosts can be prepared in a number of ways. Perhaps the most common techniques are iso-osmotic lysis and hypotonic lysis.

Protocol 13. Preparation of human erythrocyte ghost membranes by hypo-osmotic shock[a]

Equipment and reagents

- Lithium-heparinized collection tubes (Bibby Sterilin Ltd.)
- Erythrocyte wash buffer: 2 mM Hepes, 154 mM NaCl, pH 7.1 at 4°C
- Lysis buffer: 10 mM Tris-HCl, 1 mM EDTA, pH 7.1 at 4°C

Method

1. Collect fresh human blood in heparinized collection tubes.
2. Centrifuge the blood at 500 × *g* for 10 min.
3. Discard the supernatant and the 'buffy coat' of white cells.
4. Resuspend the pellet in 5 volumes of ice-cold erythrocyte wash buffer[b] and centrifuge at 500 × *g* for 10 min.
5. Repeat step 4.
6. Pour the loose pellet into 15 volumes of ice-cold lysis buffer and leave on ice for 30 min.
7. Centrifuge the suspension of lysed cells at 20 000 × *g* for 15 min.
8. Discard the supernatant and gently resuspend the pink pellet by swirling, leaving behind the denser dark pellet of unlysed cells.
9. Transfer the pink pellet to a second centrifuge tube and repeat step 7.
10. Repeat the washing procedure (steps 7–10) a further 4 times[c].

Protocol 13. *Continued*

11. Resuspend the washed ghosts in 10 mM Tris–HCl buffer, pH 7.1, without EDTA and centrifuge at 20 000 × *g* for 15 min.

12. Resuspend the packed ghosts with 1 volume of EDTA-free buffer (step 11).

13. Measure the protein content and adjust to a protein concentration of approx. 4 mg ml^{-1}.

14. Dispense 0.2 ml aliquots into microcentrifuge tubes and store at −70 °C.

[a] Method of Stephens and Downes (26).
[b] All subsequent steps are performed on ice to prevent resealing of the lysed cells.
[c] It is not necessary to remove the last traces of haemoglobin, though a penultimate wash in 10 mM Tris-HCl, 154 mM NaCl, pH 7.1 may help.

Erythrocyte ghosts prepared according to *Protocol 13* retain the Ins(1,4,5)P_3/ Ins(1,3,4,5)P_4 5-phosphatase and the multiple inositol polyphosphate phosphatase (MIPP) that removes the 3-phosphate from Ins(1,3,4,5)P_4 (54, 55). By manipulation of the ionic conditions, particularly the Mg^{2+} concentration, it is possible to modulate the activities of the enzymes underlying the observed activities (2, 54). Thus, in the absence of Mg^{2+} and the presence of EDTA, the Ins(1,4,5)P_3/Ins(1,3,4,5)P_4 5-phosphatase is totally inactive such that Ins(1,4,5)P_3 is the sole product of dephosphorylation of Ins(1,3,4,5)P_4 (54). The Mg^{2+} requirement of the 5-phosphatase activity was exploited to determine the route of synthesis of PtdIns(3,4,5)P_3 in neutrophils (2). Cells were labelled with [^{32}P]orthophosphate under non-equilibrium conditions. Ins(1,3,4,5)P_4, obtained by deacylation of PtdIns(3,4,5)P_3 and subsequent deglyceration of the GroPIns(3,4,5)P_3 product, was dissected using erythrocyte ghosts under conditions of either high (10 mM) Mg^{2+} concentration or in its absence. In the former situation the labelling of the 5-phosphate was determined and in the latter the labelling of the 3-phosphate was obtained. The same conditions employed in *Protocols 14* and *15*, detailed below, have also facilitated descriptions of the routes of synthesis of PtdIns(3,4)P_2 and PtdIns(4,5)P_2 in plant cells (8) and substrate-cycling of these components (16, 17).

(ii) Estimation of the labelling of the 5-phosphate of PtdIns(4,5)P_2

The erythrocyte ghost Ins(1,4,5)P_3/Ins(1,3,4,5)P_4 5-phosphatase activity has been shown (2, 8, 16, 17, 52) to attack GroPIns(4,5)P_2 yielding GroPIns4P and P$_i$. The products can be resolved on Partisphere SAX HPLC columns. The inclusion during dissection of a standard of [^{3}H]GroPIns(4,5)P_2, prepared by deacylation of [^{3}H]PtdIns(4,5)P_2, facilitates estimation of labelling of the 5-phosphate by a simple comparison of the ^{32}P:^{3}H ratio of the GroPIns4P product and the starting material even in the presence of other contaminating enzyme activities. These activities, which may include phospho-diesterases (56) and phosphomonoesterases, are usually present at low levels in the assay. In the absence of a ^3H standard, and providing that the contaminating activities are negligible, comparison of the radioactivity recovered in peaks with the characteristic retention times of GroPIns4P and P_i yield the label in the 1- and 4-phosphates combined and the 5-phosphate respectively. Under conditions of short-term labelling, typically 1 h, the labelling of the diester (1-phosphate) is usually negligible.

Protocol 14. Estimation of labelling of the 5-phosphate of PtdIns(4,5)P_2[a]

Equipment and reagents
- Thin-stemmed long-reach pH electrode (Type CMAWL S7/3.7/180, Russell pH Ltd.)
- Assay buffer: 12.5 mM Hepes, 10 mM $MgCL_2$, 1 mM EGTA, pH 7.2 at 25°C
- Erythrocyte ghosts (*Protocol 13*)
- 70% v/v perchloric acid

Method

1. Take up the desalted [^{32}P]GroPIns(4,5)P_2 and [^{3}H]GroPIns(4,5)P_2[a] in 0.2–0.3 ml of buffer and transfer the sample to a microcentrifuge tube.

2. Measure the pH and if necessary adjust to pH 7.2 by the addition of 2 M KOH a few μl at a time.

3. Resuspend the thawed erythrocyte ghosts in assay buffer and centrifuge the ghosts at full speed in a microcentrifuge for 5 min.

4. Discard the supernatant and repeat step 3 twice.

5. Add the [^{3}H/^{32}P]GroPIns(4,5)P_2 sample in buffer to the erythrocyte ghosts (final concentration approx. 2 mg ml^{-1} protein) and incubate the sample at 37°C for 2 h.

6. Add 10 μl of perchloric acid and hold the sample on ice for 15 min.

7. Centrifuge the sample at full speed in a microcentrifuge for 5 min at 4°C.

8. Remove the supernatant and retain it for HPLC.

9. Resuspend the pellet in 0.5 ml of ice-cold water and repeat step 7.

10. Pool the supernatants from steps 7 and 9 and perform HPLC according to *Protocol 6*.

[a]Assay conditions derived from Stephens *et al.* (2).

(iii) Estimation of the labelling of the 3-phosphate of PtdIns(3,4)P_2

The erythrocyte ghosts described in section 7.2.2 and incubated under the conditions described in *Protocol 15* attack GroPIns(3,4)P_2 with a

3-phosphatase yielding GroPIns4P and P_i only, which are easily resolved on Partisphere SAX HPLC (8, 16, 17). As for all the dissection strategies described in this chapter, the inclusion during dissection of a tritiated standard is highly desirable. However, [^{3}H]PtdIns(3,4)P_2, from which [^{3}H]GroPIns(3,4)P_2 is obtained, is not available commercially and is obtained only in very low yield from [^{3}H]Ins- labelled plant tissues. Fortunately, GroPIns4P and P_i are the only products of GroPIns(3,4)P_2 dephosphorylation under the incubation conditions described in *Protocol 15*. Thus, providing that it can be shown that the GroPIns*P* product elutes after a standard of [^{3}H]GroPIns3P and with GroPIns4P then a comparison of the ^{32}P radioactivity associated with the P_i and GroPIns4P peaks respectively reveals the labelling of the 3 phosphate and the 1 and 4 phosphates combined.

Protocol 15. Estimation of labelling of the 3-phosphate of PtdIns(3,4)P_2[a]

Equipment and reagents

- Thin-stemmed long-reach pH electrode (Type CMAWL S7/3.7/180, Russell pH Ltd.)
- Assay buffer: 12.5 mM Hepes, 5 mM EDTA, 1 mM EGTA, pH 7.2 at 25°C
- Erythrocyte ghosts (*Protocol 13*)
- 70% v/v perchloric acid

Method

1. Take up the desalted [^{32}P]GroPIns(3,4)P_2 in 0.2–0.3 ml of buffer and transfer the sample to a microcentrifuge tube.

2. Measure the pH and if necessary adjust to pH 7.2 by the addition of 2 M KOH a few μl at a time.

3. Resuspend the thawed erythrocyte ghosts in assay buffer and centrifuge the ghosts at full speed in a microcentrifuge for 5 min.

4. Discard the supernatant and repeat step 3 two times.

5. Add the [^{32}P]GroPIns(3,4)P_2 sample in buffer to the erythrocyte ghosts (final concentration approx. 2 mg ml^{-1} protein) and incubate the sample at 37°C for 4 h.

6. Process the sample and perform HPLC according to *Protocols 14* and *16*.

[a] Assay conditions derived from Stephens *et al.* (2).

Acknowledgements

We thank Sue J. Green of the Department of Plant Sciences for the culture of the plant material on which the methods described in this chapter were tested.

References

1. Irvine, R.F. (1990). In *Inositol metabolism in plants* (ed. D.J. Morré, F.A. Loewus, and W.F. Boss). Wiley-Liss, New York.
2. Stephens, L.R., Hughes, K.T., and Irvine, R.F. (1991). *Nature* (*London*), **351**, 33.
3. Stephens, L.R., Jackson, T.R., and Hawkins, P.T. (1993). *Biochim. Biophys. Acta*, **1179**, 27.
4. Carter, A.N., Huang, R., Sorisky, A., Downes, C.P., and Rittenhouse, S.E. (1994). *Biochem. J.*, **301**, 415.
5. Hawkins, P.T., Jackson, T.R., and Stephens, L.R. (1992). *Nature* (*London*), **358**, 157.
6. Corey, S., Eguinoa, A., Puyana-Theall, K., Bolen, J.B., Cantley, L., Mollinedo, F., Jackson, T.R., Hawkins, P.T., and Stephens, L.R. (1993). *EMBO J.*, **12**, 2681.
7. Woscholski, R., Waterfield, M.D., and Parker, P.J. (1996). *J. Biol. Chem.*, **270**, 31 001.
8. Brearley, C.A. and Hanke, D.E. (1993). *Biochem. J.*, **290**, 145.
9. Parmar, P.N. and Brearley, C.A. (1993). *Plant J.*, **4**, 255.
10. Graziani, A., Ling, L.E., Endemann, G., Carpenter, C.L., and Cantley, L.C. (1992). *Biochem. J.*, **284**, 39.
11. Stephens, L., Cooke, F.T., Walters, R., Jackson, T., Volinia, S., Gout, I., Waterfield, M.D., and Hawkins, P.T. (1994). *Current Biol.*, **4**, 203.
12. Kodaki, T., Woscholski, R., Emr, S., Waterfield, M., Nurse, P., and Parker, P. (1994). *Eur. J. Biochem.*, **219**, 775.
13. Volinia, S., Dhand, R., Vanhaesebroeck, B., MacDougall, L.K., Stein, R., Zvelebil, M.J., Domin, J., Panaretou, C., and Waterfield, M.D.(1995). *EMBO J.*, **14**, 3339.
14. Hong, Z. and Verma, D.-P.S. (1994). *Proc. Natl. Acad. Sci. USA*, **91**, 9617.
15. Welters, P., Takegawa, K., Emr, S.D., and Chrispeels, M.J. (1994). *Proc. Natl. Acad. Sci. USA.*, **91**, 11 398.
16. Parmar, P.N. and Brearley, C.A. (1995). *Plant J.*, **8**, 425.
17. Brearley, C.A. and Hanke, D.E. (1995). *Biochem. J.*, **311**, 1001.
18. Müller, E., Hegewald, H., Jaroszewicz, K., Cumme, G.A., Hoppe, H., and Fründer, H. (1986). *Biochem. J.*, **235**, 775.
19. King, C.E., Stephens, L.R., Hawkins, P.T., Guy, G.R., and Michell, R.H. (1987). *Biochem. J.*, **244**, 209.
20. Stephens, L.R., Jackson, T.R., and Hawkins, P.T. (1993). *Biochem. J.*, **296**, 481.
21. Harden, T.K., Hawkins, P.T., Stephens, L.R., Boyer, J.L., and Downes, C.P. (1988). *Biochem. J.*, **252**, 583.
22. Lakin-Thomas, P.L. (1993). *Biochem. J.*, **292**, 805.
23. Lips, D.L., Majerus, P.W., Gorga, F.R., Young, A.T., and Benjamin, T.L. (1989). *J. Biol. Chem.*, **264**, 8759.
24. Serunian, L.A., Haber, M.T., Fukui, T., Kim, J.W., Rhee, S.G., Lowenstein, J.M., and Cantley, L.C. (1989). *J. Biol. Chem.*, **264**, 17 809.
25. Michell, R.H., Kirk, C.J., Maccallum, S.H., and Hunt, P.A. (1988). *Phil. Trans. R. Soc. Lond.*, B **320**, 239.
26. Stephens, L.R. and Downes, C.P. (1990). *Biochem. J.* **265**, 435.
27. Jackson, T.R., Stephens, L.R., and Hawkins, P.T. (1992). *J. Biol. Chem.*, **267**, 16 627.

28. Stephens, L.R., Jackson, T.R., and Hawkins, P.T. (1993) *J. Biol. Chem.*, **268**, 17612.
29. Munnik, T., Irvine, R.F., and Musgrave, A. (1994). *Biochem. J.*, **298**, 269.
30. Drøbak, B.K. and Watkins, P.A.C. (1994). *Biochim. Biophys. Res. Commun.*, **205**, 739.
31. Stuart, J.A., Anderson, K.L., French, P.J., Kirk, C.J., and Michell, R.H. (1995). *Biochem J.*, **303**, 517.
32. Fiskum, G. (1985). *Cell Calcium*, **6**, 25.
33. Lieber, M.R. and Steck, T.L. (1982). *J. Biol. Chem.*, **257**, 11651.
34. Lieber, M.R. and Steck, T.L. (1982). *J. Biol. Chem.*, **257**, 11660.
35. Billah, M.M., Finean, J.B., Coleman, R., and Michell, R.H. (1976). *Biochim. Biophys. Acta*, **433**, 54.
36. Munnik, T., Arisz, S.A., de Vrije, T., and Musgrave, A. (1996). *Plant Cell.*, **7**, 2197.
37. Galliard, T. (1980). In *Biochemistry of plants* , Vol. 4 (ed. P.K. Stumpf), p. 85. Academic Press, New York.
38. Scherer, G.F.E. and Morré, D.J. (1978). *Plant Physiol.*, **62**, 933.
39. Wreggett, K.A. and Irvine, R.F. (1987). *Biochem. J.*, **245**, 655.
40. Boss, W.F. (1989). In *Second messengers in plant growth and development* (ed. W.F. Boss. and D.J. Morré), pp. 29–56. Alan R Liss, New York.
41. Irvine, R.F., Letcher, A.J., Lander, D.J., Drobak, B.K., Dawson, A.P., and Musgrave, A. (1989). *Plant Physiol.*, **89**, 888.
42. Wissing, J.B. and Behrbohm, H. (1993). *FEBS Lett.*, **315**, 95.
43. Wissing, J.B. and Behrbohm, H. (1993). *Plant Physiol.*, **102**, 1243.
44. Clarke, N.G. and Dawson, R.M.C. (1981). *Biochem. J.*, **195**, 301.
45. Hawkins, P.T., Stephens, L.R., and Downes, C.P. (1986). *Biochem. J.*, **238**, 507.
46. Stephens, L.R., Hawkins, P.T., and Downes, C.P. (1989). *Biochem. J.*, **259**, 267.
47. Brown, D.M. and Stewart, J.C. (1966). *Biochim. Biophys. Acta*, **125**, 413.
48. Whitman, M., Downes, C.P., Keeler, M., Keller, T., and Cantley, L. (1988). *Nature* (London), **332**, 644.
49. Wreggett, K.A. and Irvine, R.F. (1989). *Biochem. J.*, **262**, 997.
50. Stephens, L.R., Hawkins, P.T., Barker, C.J., and Downes, C.P. (1988). *Biochem. J.*, **253**, 721.
51. Stephens, L.R. (1990). In *Methods in inositide research* (ed. R.F. Irvine), pp. 9–29. Raven Press, New York.
52. Hawkins, P.T., Michell, R.H., and Kirk, C.J. (1984). *Biochem. J.*, **218**, 785.
53. Shears, S.B. (1989). *Biochem. J.*, **260**, 313.
54. Estrada-Garcia, T., Craxton, A., Kirk, C.J., and Michell, R.H. (1991). *Proc. R. Soc. Lond.*, B **244**, 63.
55. Craxton, A., Ali, N., and Shears, S.B. (1995) *Biochem. J.*, **305**, 491.
56. Downes, C.P. and Michell, R.H. (1981). *Biochem. J.*, **198**, 133.

Measurement of inositol phosphate turnover in intact cells and cell-free systems

STEPHEN B. SHEARS

1. The assay of inositol phosphate accumulation by batch elution from anion-exchange columns as an index of the rate of PLC activity

1.1 Why this assay?

Studies into the agonist-mediated regulation of inositol lipid hydrolysis by PLC represent a cornerstone of signal transduction research (1). There are three potential lipid substrates: PtdIns, PtdIns$P$ and PtdIns(4,5)P_2. It is PtdIns(4,5)P_2 that receives the most attention as far as receptor-regulated PLC activity is concerned, because both of the immediate reaction products, Ins(1,4,5)P_3 and 1,2-diacylglycerol, are intracellular signals (1). The extent to which PLC might also attack PtdIns4P and PtdIns is usually uncertain, and few studies have successfully answered this difficult question. One notable exception used pancreatic islets which were stimulated with carbachol. PtdIns(4,5)P_2 hydrolysis was thought to predominate during the initial few minutes of stimulation (2). However, after one hour of stimulation there was evidence that the rate of PtdIns hydrolysis by PLC may have exceeded that of PtdIns(4,5)P_2 (2). The significance of PtdIns as a substrate probably relates to it being a supplementary source of 1,2-diacylglycerol.

The question as to whether a particular stimulus activates PLC activity in intact cells can be answered by demonstrating either that levels of the inositol lipid substrates decrease (see Chapter 1), or that there are increases in the levels of Ins(1,4,5)P_3 (see below and Chapter 8) and/or 1,2-diacylglycerol. However, more detailed knowledge of the *rate* of PLC activity will not be obtained from these specific assays alone. For example, one of the consequences of receptor occupation is a stimulation of inositol lipid resynthesis to ensure a compensatory supply of substrate to ensure continuing PLC activity (3). This is why changes in inositol lipid levels are not a reliable measure of

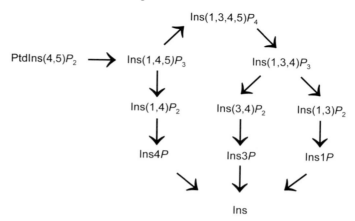

Figure 1. The pathway of conversion of Ins(1,4,5)P_3 to free inositol, as derived mainly from data obtained from experiments on animal cells. Some of these metabolites are representatives of alternate metabolic pathways in, for example, higher plants (32, 46), algae (10), yeast (43) and *Dictyostelium* (44, 45). Note that Ins(1,4)P_2 and Ins1P may also be formed by direct PLC-mediated hydrolysis of PtdIns4P and PtdIns.

the rate of PLC activity. As for assaying the products of PtdIns(4,5)P_2 hydrolysis—Ins(1,4,5)P_3 and 1,2-diacylglycerol—these are rapidly metabolized. An estimate of the rate of PLC activity is only reliable if the levels of the downstream metabolites of these two intracellular signals can also be measured. This is not practicable as far as diacylglycerol is concerned, because of the difficulties involved in accurately accounting for the extent of both its deacylation, and its phosphorylation and reincorporation back into inositol lipids (4). Furthermore, diacylglycerol release by the action of PLC is supplemented by hydrolysis of phosphatidylcholine, particularly during prolonged cell activation (4). It is difficult to distinguish between the two lipid sources of diacylglycerol.

Fortunately, the characteristics of Ins(1,4,5)P_3 metabolism are more amenable to experimental manipulations that make it possible to use inositol phosphate turnover as an index of receptor-dependent PLC activity. Within the time-frame that these phospholipase assays are usually performed (< 1 h), the Ins(1,4,5)P_3 that is released inside animal cells is metabolized primarily by the pathway shown in *Figure 1*. Thus, the basis for the methodology described in this section relies on specifically incorporating radiolabel into the inositol lipids and using anion-exchange chromatography to record stimulus-dependent changes in the levels of the metabolites Ins$P$, InsP_2, InsP_3, and InsP_4. This can be done by HPLC—the methodology for this is discussed in section 2—but each tissue sample has then to be processed individually and relatively slowly. Alternately, a batch procedure can be used. The major advantage of this technique is that a large quantity of samples can be processed quite rapidly, so that many useful data on PLC regulation will accumulate in a relatively short time.

1.2 Labelling cells with [³H]inositol

Cellular inositol phosphates can be labelled with [³²P], and such studies have provided very useful information (see e.g. refs. 5, 6). However, this radio-isotope gets incorporated into many other organic phosphates which makes it difficult to assay routinely the specific turnover of [³²P]-labelled inositol phos-phates. Thus, the accumulation of inositol phosphates following PLC activation is generally measured after pre-incubating tissue slices, or cell preparations, with either [2-³H]- or [U-¹⁴C]-labelled inositol. [³H]inositol is usually pre-ferred because it is far less expensive. In animal cells, around 5% of added radioactivity can be expected to accumulate into the inositol lipids at isotopic equilibrium under optimal conditions (7). The level of incorporation into cell preparations from higher plants is considerably lower (8). Some lower plants, such as *Chlamydomonas*, are also very difficult to label with [³H]inositol (9, 10).

It is axiomatic that changes in levels of [³H]inositol phosphates are only unequivocally reflective of changes in their mass levels if the relevant inositol lipid precursors, and the inositol phosphate products, are all labelled to iso-topic equilibrium. For the PLC-accessible pools of PtdIns(4,5)P_2, and the inositol phosphate products depicted in *Figure 1*, equilibrium labelling is nor-mally attained within 48 h of incubation with [³H]inositol (11, 12). The approach to a steady-state labelling condition can be easily monitored by per-forming a time course of [³H]inositol accumulation into the inositol lipids (see *Protocol 1*). Nevertheless, it is best to prove that equilibrium labelling of the Ptdins(4,5)P_2 has actually been attained; to obtain this information the experimenter should determine the specific radioactivity of this pool by performing the mass assays described in Chapter 8.

During the radiolabelling protocol, the concentration of non-radioactive inositol added to the incubation medium is an important factor to be taken into consideration. This can not only have an impact on the goal of labelling to isotopic equilibrium, but it is also relevant to the objective of maintaining adequate mass levels of the inositol lipids (11, 12). The exact protocol for achieving these requirements has to be determined empirically due to vari-ability between cell types in the rate of inositol uptake, and to differences in inositol lipid synthesis. To illustrate the range of inositol concentrations that might be necessary, two studies with different types of animal cells found that 50 μM (11) or 400 μM (12) were needed. For those organisms that can be grown on both liquid and solid media (the filamentous fungus *Neurospora crassa*, for example) the amount of inositol needed can be different in each case (13). As for the amount of radioactivity, typically 10–20 μCi ml⁻¹ [³H]inositol can be used successfully. In animal cells at least, there does not seem to be significant exchange of label out of the inositol moiety within the time-frame of the labelling protocol.

The necessity to label cells for 48 h or more means that [^{3}H]inositol is usually added to the cells in their favourite culture medium, which may contain factors that themselves affect PLC activity. The culture medium is therefore replaced with serum-free medium once the labelling period is complete. Since the new buffer need not contain any [^{3}H]inositol, this procedure also has the advantage of removing substantial unwanted radioactivity from the incubations.

1.3 Use of lithium in PLC assays

The Ins(1,4,5)P_3 that is formed by PLC-dependent PtdIns(4,5)P_2 hydrolysis is rapidly metabolized, both by a 5-phosphatase to Ins(1,4)P_2 and by a 3-kinase to Ins(1,3,4,5)P_4 (*Figure 1*). These metabolites are themselves rapidly dephosphorylated, and any that are hydrolysed to free inositol—which occurs quite quickly—represents a loss of products from the assay (because the [^{3}H]inositol peak is so relatively huge it is not practical to assess the relatively small changes in its levels that might occur during PLC stimulation, and in any case the [^{3}H]inositol is utilized for lipid resynthesis). It is therefore a common practice to *reduce* the metabolic flux from inositol polyphosphates to free inositol by using lithium as a metabolic trap. The italicized emphasis in the last sentence was carefully chosen. Lithium inhibits Ins(1,4)P_2/Ins(1,3,4)P_3 1-phosphatase and inositol monophosphatase (14), but it does so in an uncompetitive manner. Therefore the efficacy of lithium as an inhibitor is critically dependent upon substrate concentration (14). It is sometimes forgotton that there can be a metabolic 'leak' past the lithium block to free inositol. The extent of the 'leak' in the *stimulated* condition can be determined by first reversing the action of the agonist with an antagonist, and then monitoring the time course of the decay in inositol phosphate levels. Under basal conditions there are lower cellular levels of inositol phosphates, so the degree of inhibition by lithium in this condition will also be lower. Nevertheless, lithium may still increase basal levels of inositol phosphates, albeit relatively slowly. This means that the base-line for the controls can change during the time course of the experiment, so it is important to compare the *time course* for changes to inositol phosphate levels in both resting and stimulated cells.

There is another lithium-insensitive pathway by which [^{3}H]inositol label can escape the metabolic cycle depicted in *Figure 1*: phosphorylation of [^{3}H]Ins(1,3,4)P_3 to [^{3}H]Ins(1,3,4,6)P_4 (*Figure 2*). The [^{3}H]Ins(1,3,4,6)P_4 will be accounted for in the assay of PLC activity (since it includes all of the [^{3}H]InsP_4). There is evidence that the specific activity of [^{3}H]Ins(1,3,4,6)P_4 can change during PLC activation (15), but this will not detectably impact on the accuracy of the PLC assays, since [^{3}H]Ins(1,3,4,6)P_4 is such a minor constituent of the total (InsP_1 + InsP_2 + InsP_3 + InsP_4).

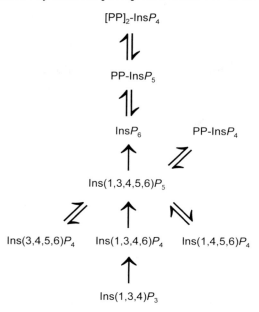

Figure 2. A pathway to the higher inositol polyphosphates. The metabolites depicted in this metabolic pathway are, in animal cells at least, the subject of active investigation because there is growing evidence that they are important in cell signalling. Other intermediates, not listed here, participate in further important pathways of $InsP_6$ synthesis and metabolism in *Dictyostelium* (44), higher plants (32, 46) and even in animal cells (33).

Different cells accumulate lithium at different rates and to varying extents. It is therefore a good idea to determine empirically the lithium treatment that leads to the most complete inhibition of inositol phosphate degradation (as discussed above, this can be achieved by monitoring the rate of loss of [^{3}H]inositol phosphates after termination of the extracellular stimulus). It is usual to add to incubations a final concentration of 10–20 mM lithium (as LiCl), some 10–20 minutes before the experimental protocol begins. Higher concentrations tend to be avoided because they increase the opportunity for non-specific effects of this cation impacting upon the response being measured. Note that lithium is of less practical use in higher plant systems, where many pathways of inositol phosphate dephosphorylation show either poor or no sensitivity towards lithium (16).

Note that by virtue of inhibiting inositol phosphate dephosphorylation, lithium can cause the supply of [^{3}H]inositol to become insufficient to support the rate of inositol lipid resynthesis necessary to sustain a prolonged agonist stimulation; the consequence of this is attenuation of PLC activity (11). The problem can largely be avoided by maintaining the mass levels of inositol lipids at their maximum levels during the labelling period. As mentioned above (section 1.2), this is achieved by empirically optimizing the supply of inositol mass in the culture medium.

1.4 The time course of PLC assays

PLC activity during the experiment is calculated from the difference in the time course of inositol phosphate accumulation in the presence and absence of an appropriate stimulus. It should be noted that the time course for the stimulated condition is frequently biphasic—a relatively rapid initial phase is followed by a slower secondary phase (11). In some cases the initial phase is less than 30 sec in duration (17). The precise pattern will depend upon the time course and the extent to which a receptor undergoes homologous de-sensitization (11, 17). Analysis of both of these phases will require a judicious choice of time points in these experiments.

Another factor that is worth considering before designing a time course is the possibility that in some cell types PLC may also attack PtdIns and PtdIns4P (see e.g. ref . 2). Any Ins1P and Ins(1,4)P_2 released in this manner will of course be accounted for in the assay of released inositol phosphates, but it is worth remembering that total PLC activity may not be equivalent to net PtdIns(4,5)P_2 hydrolysis. The relative contributions from hydrolysis of PtdIns and PtdIns4P may also change with time of stimulation (2).

1.5 The methodology for quenching intact cell incubations and extracting the soluble inositol phosphates

The investigator must also chose a rapid method to quench the incubations. A number of quench media have been employed, but some have potential pitfalls. Media that are based on a mixture of chloroform and methanol should be avoided since there can sometimes be problems with recovery of inositol polyphosphates (18), and it can also lead to the non-physiological production of methyl-Ins(1,4,5)P_3 (19). Therefore, incubations are most frequently quenched with either trichloroacetic acid or perchloric acid. Most inositol phosphates are acid-stable, except for those with a 1,2-cyclic diester; these compounds represent a small proportion of the products of PLC. An acid-quenched sample that originally contained Ins(1:2cyclic)P for example, will have this compound converted to Ins2P plus Ins1P, although of course these will still be included in the PLC assay. A complication only really arises if levels of individual inositol phosphates are being assayed. Those workers who are interested in quantifying cyclic inositol phosphates should quench with boiling water (20), or adopt a neutral phenol/chloroform quench medium (21); special HPLC separation procedures are also required to resolve cyclic inositol phosphates from their non-cyclic counterparts (21).

The acid-quenched samples must subsequently be neutralized otherwise the acid will react with the chromatographic resins that are used to measure inositol phosphates. Trichloroacetic-quenched samples are frequently neutralized by repeated washes with diethyl ether (in which the acid is soluble). Despite the widespread use of this method it is, in the author's opinion, a laborious procedure, and in any case the use of diethyl ether is a safety hazard. A mixture of freon and octylamine can also be used to neutralize cell extracts, but this procedure is not recommended because of its environmental impact. Instead, samples can be safely and conveniently quenched with perchloric acid, and neutralized with K_2CO_3. This method is described in *Protocol 1*.

Protocol 1. Quenching incubations of [^{3}H]inositol-labelled cells for analysis of inositol phosphates

Equipment and reagents
- Bench microcentrifuge
- Vacuum pump
- 100 ml 0.6 M perchloric acid plus 0.2 mg ml^{-1} InsP_6
- 100 mM NaOH plus 0.1% (v/v) Triton X-100
- 100 ml 1 M K_2CO_3 plus 5 mM EDTA
- 10 ml Universal Indicator

Method

1. If cells are attached to a culture plate, rapidly apirate off the incubation medium using a vacuum line and quench with 1.5 ml^a of 0.6M perchloric acid plus 0.2 mg ml^{-1} Ins$P_6{}^b$. If cells are in suspension, quench the incubation with 0.4 volumes of 2 M perchloric acid plus 1 mg ml^{-1} InsP_6.

2. Place quenched samples on ice for 15 min.

3. Remove acidified sample, place in a polypropylene tube and neutralize (see step 5).

4. The quantities of [^{3}H]inositol phosphates that are obtained for each sample are normalized so as to compensate for any sampling variation. The level of [^{3}H] incorporated into total inositol lipids is a good common denominator (see *Protocol 2*). Thus, the lipids are extracted from the cell debris that remains attached to the culture plates, as follows:

    (a) Wash twice with 1.5ml of ice-cold 0.6 M (v/v) solution of perchloric acid.

    (b) Dissolve the lipids in 1.5 ml 100 mM NaOH plus 0.1% (v/v) Triton X-100.

    (c) Store samples overnight at 0–4°C.

    (d) Count aliquots for radioactivity the next day.

5. To the supernatants obtained in step 3, neutralize with an appropriate volume of 1M K_2CO_3, 5 mM EDTA^c (the exact volume required can be determined by preparing 'blank' extracts supplemented with 10–20 µl Universal Indicator, which will turn green at neutral pH). Samples

---

Protocol 1. *Continued*

should be kept on ice for at least 2 h (overnight storage is preferable)^d. During this time the sample tubes should only be loosely capped, so as to facilitate release of CO_2. The precipitate is removed by centrifugation at either 10 000 g for 2 min, or 2500 g for 15 min, at a temperature of 0–4°C. The supernatant is saved for chromatographic analysis (*Protocol 2*).

^{a}This protocol assumes that cells have been cultured in 35 mm wells; all volumes can be adjusted proportionately if different sized wells were used.
^{b}To aid recovery of [^{3}H]inositol phosphates.
^{c}The EDTA chelates cations and prevents them from causing precipitation and loss of inositol phosphates.
^{d}To facilitate the precipitation of a poorly soluble potassium perchlorate.

1.6 Assay for inositol phosphates using gravity-fed ion-exchange columns

The PLC assay requires measurement of the levels of [^{3}H]Ins$P$ + [^{3}H]InsP_2 + [^{3}H]InsP_3 + [^{3}H]InsP_4 in a quenched and neutralized extract. These polyphosphates must therefore be separated from [^{3}H]inositol, [^{3}H]glycerophosphoinositol, and the [^{3}H]InsP_5 + [^{3}H]InsP_6 that is also present. This can be achieved by batch elution from gravity-fed anion exchange chromatography columns (*Protocol 2*).

Protocol 2. Batch elution of inositol phosphates on gravity-fed columns

Equipment and reagents

- 500 g ion-exchange resin, AG 1-X8 200–400 mesh, formate form (Bio-Rad)
- 100 Polyprep propylene column supports (Bio-Rad)
- 1 l 0.06 M ammonium formate
- A plexiglass (perspex) rack to accommodate the columns
- 1 l 1.05 M ammonium formate/0.1 M formic acid

Method

1. Wash the ion-exchange resin (which is supplied in a dry form) twice with distilled water, and remove any 'fines' that float to the surface. (Store the resin at 0–4°C in a dark bottle. Some batches of resin are particularly prone to turning green when exposed to sunlight, but this can be avoided by covering the beaker with foil when the resin is washed.)

2. Stir a slurry of resin (1:1.3 wet wt. resin:water) continuously and as fast as possible to avoid settling of the resin. When pipetting the slurry, use a tip with the last few millimeters cut from the end, to prevent a blockage forming. Apply approx. 5 ml water to a column sup-

port[a] followed by sufficient resin for a final column size of 0.8 ml packed volume[b]. Allow all the water to drain.

3. Check all columns are the same size.

4. To a single column add 9 ml water to reservoir, then immediately add neutralized sample. Repeat for each sample. [³H]inositol will elute in the flow-through and can usually be discarded. For this step, and during all subsequent elutions, the columns are gravity-fed.

5. Elute residual [³H]inositol with 2 × 5ml aliquots of water[c]. Minimize disturbance to the resin bed by letting the water flow down the side of the reservoir of the column support.

6. Elute [³H]glycerophosphoinositol with 10 ml 0.06 M ammonium formate. The eluate is also discarded.

7. Elute [³H]InsP_1, [³H]InsP_2, [³H]InsP_3 and [³H]InsP_4 all together[d] with 2 × 5ml aliquots of 1.05 M ammonium formate/0.1 M formic acid. Save the eluate, take an aliquot and mix with scintillant to assay [³H].

8. For PLC assays, the quantities of [³H]inositol phosphates that are obtained for each sample are frequently normalized so as to compensate for any sampling variation. The level of [³H] incorporated into total inositol lipids is a good common denominator. In other words, the levels of each [³H]inositol phosphate are expressed as a percentage with reference to radioactivity in the lipids determined for the same sample (see *Protocol 1*).

[a]The addition of water first to the column support helps to prevent air-pockets forming as the resin settles.
[b]The suggested elution protocol must be checked with standards, and it may prove necessary to alter the column size and the volume or concentration of eluates to maximize the column resolution.
[c]The free [³H]inositol must be completely removed, because it is present in such large quantities relative to the levels of [³H]inositol phosphates.
[d]InsP_5 and InsP_6 will not usually be eluted from the column by this protocol, but this fact should be checked with standards.

1.7 Validating that this batch analysis is a reliable assay of PLC activity

Three distinct studies illustrate the importance of confirming that the data obtained from batch analyses (*Protocol 2*) really do reflect genuine PLC activity. First, it has been shown that the stimulation of cultures of [³H]inositol-labelled *Saccharomyces cerevisiae* with glucose causes a release of [³H]GroPIns, [³H]GroPIns$P$ and [³H]GroPInsP_2 without a significant production of inositol phosphates (22). The authors established this fact by analyzing the structures of the [³H]-labelled compounds by HPLC (22). Although GroPIns is excluded from the batch analysis (see above), GroPIns$P$ and GroPInsP_2 are not (their levels, at least in animal cells, are generally too low to be of concern). Therefore if in such a study the accumulation of soluble [³H]-labelled compounds had been assayed by this batch analysis alone, it could have been erroneously concluded that PLC was glucose-activated.

A higher plant system provides the second example of the difficulties that can arise with this batch analysis. Plants metabolize inositol to cell-wall constituents such as arabinose, xylose and galacturonic acid. These and other compounds probably contribute to the high background of [³H]-labelled material that swamps the much smaller levels of [³H]inositol phosphates that are present (16, 23).

Animal cells may also present problems on occasions: recently a study on FRTL-5 thyrocytes challenged with thyroid-stimulating hormone (TSH) was reported (24). There were no ensuing changes in the levels of either inositol lipids, nor in $Ins(1,4,5)P_3$, $Ins(1,4)P_2$ or $Ins(1,3,4,5)P_4$ (as determined by HPLC, see below). This agonist did elevate levels of an unidentified $InsP_3$ (possibly $Ins(1,3,4)P_3$) by an unexplained route, and also appeared to cause a net dephosphorylation of $Ins(1,3,4,5,6)P_5$ to $Ins(1,4,5,6)P_4$ and/or $Ins(3,4,5,6)P_4$ (24). There was therefore an agonist-dependent accumulation of $InsP_4$ and $InsP_3$, but, in the authors opinion, this was not as a consequence of activation of PLC activity. These authors pointed out that any judgment concerning the regulation of PLC activity by TSH could have been erroneous, had it been based solely on the elevation in levels of inositol polyphosphates as measured by the batch techniques described here—as indeed was the case in earlier studies of TSH action made by other laboratories. (Note that receptor-mediated conversion of $Ins(1,3,4,5,6)P_5$ to $Ins(3,4,5,6)P_4$ frequently accompanies PLC activation (15), but usually the amount of $InsP_4$ formed is too small to impact seriously on the total amount of inositol phosphates produced as a direct consequence of PLC activation).

These three studies demonstrate that it is important to support data obtained from batch analyses with additional results that typify stimulus-dependent PLC activity. These include confirmation of transient changes in levels of inositol lipids (Chapter 1) or changes in levels of $Ins(1,4,5)P_3$ and its immediate metabolites (by HPLC as described in section 2, or by the mass assays outlined in Chapter 8). Of course, the experimenter may choose to dispense with the batch methodology altogether and just use HPLC. However, despite its inadequacies and potential pitfalls, when batch analysis can be employed its value lies in it being a means of quite rapidly processing large numbers of samples simultaneously and reasonably accurately.

2. Determination of levels of individual inositol polyphosphates in intact cells

2.1 Why use HPLC assays?

The batch elution procedures described in *Protocol 2* are very useful for assessing the rate of PLC activity. However (see section 1.6) such an approach will sometimes be equivocal, and at least some HPLC analyses are useful either for validating that a given stimulus genuinely is increasing PLC activity, or as an alternative when the experimental system plainly does not lend itself to batch analysis—as is the case with higher plants, in which the [^{3}H] in [^{3}H]inositol will get incorporated into compounds other than inositol phosphates (16, 23). HPLC analyses are also necessary when InsP_6 metabolism is being studied, since the products (InsP_5, and the 'inositol pyrophosphates', see section 3) cannot readily be resolved from the InsP_6 substrate by batch analysis on gravity-fed ion exchange columns.

Frequently it is important to determine the changes in levels of individual isomers of inositol polyphosphates, particularly those with proven or putative second-messenger activities. Since the isomeric versatility of inositol phosphates is prolific—five naturally occuring isomers of InsP_4 occur in animal cells (see below and *Figures 1 and 2*)—and because such isomers cannot be separated by batch techniques, HPLC resolution again becomes obligatory. Yet even HPLC may represent only an initial step in the eventual characterization of a compound of interest. Further purification and analysis may still be necessary (see below).

2.2 HPLC analysis of inositol phosphates

A small number of laboratories directly measure the mass levels of individual inositol phosphates in a tissue extract using an on-line mass detection HPLC system (25). This methodology has failed so far to become more widely used, possibly because the sample processing and analysis is more technically demanding (see ref. 25 for details). There is also a danger of an inositol phosphate undergoing acid-catalyed phosphate migration during the analysis (26). The more accessible and most widely-used HPLC method for analyzing changes in levels of inositol phosphates is based upon their first being radiolabelled; it is this technique which is discussed here.

The factors that need to be taken into consideration during radiolabelling of cells with [^{3}H]inositol were outlined above (section 1.2), and include the need to label to isotopic equilibrium. This can usually be achieved for the inositol phosphates depicted in *Figure 1* within 48 h (see section 1). However, if it is necessary to measure the changes in levels of Ins(1,3,4,5,6)P_5, InsP_6 and their metabolites (*Figure 2*), more prolonged labelling times are obligatory. For Ins(1,3,4,5,6)P_5, 72 h is typically suitable (15). For InsP_6, even longer may be necessary—sometimes as much as 14 days (27). The approach to a steady-state radiolabelling condition can be determined empirically simply by monitoring the levels of [^{3}H]Ins(1,3,4,5,6)P_5 and [^{3}H]InsP_6. Then with the help of mass assays it can be determined whether isotopic equilibrium has been reached (Chapter 3). An alternative approach is to label cells through at least five cell-doubling times; at this point more than 96% of the cellular material present must have been synthesized from the constituents of the culture medium (28). However, this method will be compromised if cells synthesize inositol from unlabelled precursors, such as glucose-6-phosphate (29). In non-animal systems, mutants lacking pathways of endogenous inositol synthesis have proved a useful means of circumventing this difficulty (13).

The radioactive eluate from the HPLC can either be collected as discrete fractions, or it can be analysed using an on-line detector. In the latter case in particular, the amount of radioactivity used to radiolabel the cells needs to be higher than that used for gravity-fed ion-exchange columns (section 1); it is not unusual to use 50–100 μCi ml^{-1}, or even more. Such high levels of radioactivity are necessary for two reasons. First, they compensate for the low counting efficiency of some on-line scintillant/salt mixtures (typically 20–25% using the methods described in *Protocol 3*). In addition, on-line resolution of closely-eluting peaks requires short update times (we routinely use 6 sec); to reduce the extent of the statistical error during these short counting periods, a larger amount of radioactivity is required.

Although the use of lithium as a metabolic 'trap' is very helpful for

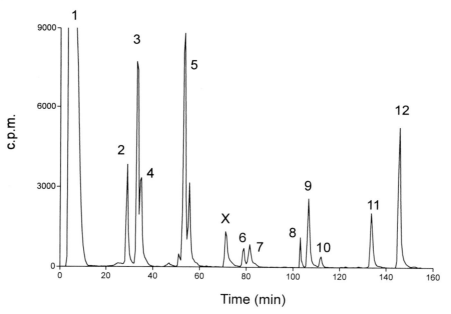

Figure 3. Adsorbosphere SAX HPLC analysis of [³H]inositol phosphates. Jurkat T-lymphocytes were labelled with 50 μCi ml^{-1} [³H]inositol for 96 h. Cells were quenched (*Protocol 1*), neutralized (*Protocol 1*) and analysed (*Protocol 3*) using an Adsorbosphere SAX column and a Packard Flo-1 on-line scintillation counter. Peak 1 = inositol; peak 2 = GroPIns; peak 3 = Ins1*P* + Ins3*P*; peak 4 = Ins4*P*; peak 5 = Ins*P*₂ (individual isomers incompletely resolved); peak 6 = Ins(1,3,4)*P*₃; peak 7 = Ins(1,4,5)*P*₃, peak 8 = Ins(1,3,4,6)*P*₄; peak 9 = Ins(1,3,4,5)*P*₄; peak 10 = Ins(3,4,5,6)*P*₄ + Ins(1,4,5,6)*P*₄. These unpublished data were kindly provided by Dr. Martina Sumner, NIEHS.

measuring the rate of PLC activity (section 1), unless this is the primary goal of the HPLC assay, lithium is usually best omitted. In HPLC assays, the main value of which is to measure physiologically relevant levels of individual inositol phosphates, lithium perturbs the very parameters being measured. For example, because lithium inhibits dephosphorylation of $Ins(1,3,4)P_3$ and the inositol monophosphates, the levels of these compounds become elevated to much higher levels that they would under physiological conditions. It has also been reported that lithium can elevate $Ins(1,4,5)P_3$ levels by about 15% through an unknown mechanism in both animal (20) and plant (30) systems.

Many laboratories have developed their own HPLC techniques, so here we will just consider that which is commonly utilized in the author's laboratory. This is based on the Adsorbosphere SAX column (*Protocol 3*). Despite the name, this column seems to exhibit properties intermediate between strong and weak anion-exchange. Although it will only partially resolve $InsP_1$ and $InsP_2$ isomers (see *Figure 3*), it does have the advantage of clearly separating many of the higher polyphosphates: $Ins(1,4,5)P_3$, $Ins(1,3,4)P_3$, $Ins(1,3,4,6)P_4$, $Ins(1,3,4,5)P_4$, $Ins(3,4,5,6)P_4/Ins(1,4,5,6)P_4$, $InsP_5$ and $InsP_6$ (*Figure 3*). In our hands this technique gives poor recovery and resolution of the newly discovered inositol pyrophosphates; an alternative dedicated HPLC procedure has been developed specifically for this purpose (section 3). Note that the separation of naturally occurring enantiomeric pairs (e.g. $Ins(3,4,5,6)P_4$ and $Ins(1,4,5,6)P_4$) can only be achieved by enantiomerically specific methodology.

Protocol 3. HPLC analysis of inositol phosphates using the Adsorbosphere SAX column

Equipment and reagents

- An HPLC system. (The minimum hardware requirement for an HPLC system is an injector, two pumps, a gradient controller, and either a fraction collector or an on-line radioactivity detector. Usually the set-up is computer-controlled. Titanium pumps have superior solvent resistance as compared with those using stainless steel.)
- Silica pre-column[a]

- 5 μ Adsorbosphere 250 × 4.6 mm SAX HPLC cartridge and holder (Alltech Associates)
- Buffer A: 1000 ml distilled, filtered and degassed water
- Buffer B: 1000 ml filtered and degassed buffer (1 M $NH_4H_2PO_4$ pH 3.35 with H_3PO_4)

Method

1. Wash out the methanol in which the Adsorbosphere columns are supplied (and in which they should also be stored) with 60 ml water (Buffer A) applied at a flow rate of 1 ml min^{-1}.

2. Wash the column with 100% buffer B for 15 min, followed by 15 min with Buffer A.

3. Construct a gradient with a flow-rate of 1 ml min^{-1} as follows (for a new column): 0–20 min, B = 0%; 20–140 min, B increases linearly to

Protocol 3. *Continued*

100%, 140–160 min, B = 100%; 161 min, B returns to 0%. There should then be a 15 min wash with water before the next run begins.

4. Pre-equilibrate the HPLC column with a 'blank' run, check the elution properties with standards, and then apply the samples.

5. Collect the eluate from the HPLC column as fractions, or assay on-line using a suitable detector and scintillant (we use Monoflow from National Diagnostics). The elution time for a particular inositol phosphate will decrease with column age (typically by approx. 2 min per 10 runs). To compensate for this, progressively decrease the maximum concentration of buffer that the column is subjected to at the end of each gradient, in 2–3% stages approximately every 10–15 runs.

6. wash the column at the end of the experiment with 30 ml water followed by 60 ml methanol (the column should be stored in the latter).

[a] The Adsorbosphere SAX HPLC column is silica based, and the silica will dissolve in water and certain buffers, causing loss of retention, increasing back-pressure, and eventually resulting in the formation of a void or channel in the column. These difficulties can be prevented by saturating the mobile phase with a silica pre-column placed between the pumps and the injector).

The identity of the compounds that are annotated in *Figure 3* relies on earlier studies showing co-chromatography with standards, and some rigorous structural identifications (e.g. ref. 15). The investigator is strongly advised to check the elution properties using commercially available radiolabelled standards. Internal standards are better, but then they must be radiolabelled with a different isotope ($[^{32}P]$ or $[^{14}C]$) and these are not readily available (except for $[^{14}C]Ins3P$).

Occasional unidentified peaks of $[^{3}H]$-labelled material are observed in HPLC runs ('X' in *Figure 3* is an example). Other unsuspected compounds will also sometimes hide within the 'known' peaks. For example, some years ago we discovered that the peak of $Ins(1,3,4,6)P_4$ extracted from the AR42J pancreatoma can also contain some $Ins(1,2,3,4)P_4$ (15). In a different study we identified $Ins(1,2,3)P_3$ in the T5–1 lymphocyte cell-line. At that time this was a novel mammalian inositol polyphosphate (31); now it appears that $Ins(1,2,3)P_3$ is a common and sometimes relatively abundant ingredient of the basal $InsP_3$ fraction in cells (32). Neither $Ins(1,2,3)P_3$ nor $Ins(1,2,3,4)P_4$ are depicted in *Figures 1 and 2*; their metabolism in animal cells still needs to be fully characterized, but they are probably constituents of pathways of $InsP_6$ synthesis and breakdown (33) as is indeed the case in plant systems (32).

It is because of these complications that when the levels of an individual inositol phosphate are of particular interest, it is useful to confirm that there are no other isomers contaminating the peak. This is an issue of special relevance when using HPLC methodology to study inositol polyphosphate turnover outside the animal kingdom, where there may be a wider range of uncertain [³H]-labelled compounds formed after an [³H]inositol labelling protocol, for example in yeast and plants. One approach to this structural problem is to isolate and desalt (Chapter 9) the polyphosphate, and then identify the [³H]polyol formed after it has been subjected to periodate oxidation, reduction and dephosphorylation (see refs. 34 and 35 for reviews of these techniques). It is also useful to check the nature of a [³H]polyphosphate by incubation with an enzyme that metabolizes it. For example, putative [³H]Ins(1,4,5)P_3 could be incubated with the Ins(1,4,5)P_3 3-kinase and the amount of [³H]Ins(1,3,4,5)P_4 formed could be measured by gravity-fed chromatography (*Protocol 5*). The presence of an internal genuine standard ([³²P]- or [¹⁴C]-labelled) is particularly useful for a quantitative analysis. Indeed, this was the approach used by our laboratory to provide the first unequivocal identification of Ins(1,4,5)P_3 in a higher plant (36). As an increasing number of the inositol phosphate metabolizing enzymes become cloned, so it is becoming easier to obtain them in a pure form. The laboratories involved can also provide insight into the optimal incubation conditions for a particular enzyme.

3. Analysis of the levels of 'inositol pyrophosphates' in intact cells

Cells have to date been shown to contain three different inositol pyrophosphates: PP-InsP_4, PP-InsP_5 (a.k.a InsP_7) and [PP]$_2$-InsP_4 (a.k.a. InsP_8) (37–39). The functions of these particular compounds are unknown, but several features suggest that important roles waits to be discovered. First, they exhibit rapid turnover in substrate cycles (37, 38). Secondly, the free-energy change when they are hydrolysed is substantial (39), and the ATP-dependent kinases that synthesize the 'inositol pyrophosphates' are physiologically sufficiently reversible so as to drive ATP synthesis (40); they may be donors in additional phosphotransferase reactions. Thirdly, recent NMR analyses of these compounds in *Dictostelium* indicate that the diphosphate groups of [PP]$_2$-InsP_4 may be in the 4- and 5- positions (41); the 4,5- vicinal monophosphates are well-known to confer specificity for a number of physiologically important actions of inositol lipids and inositol phosphates.

The PP-InsP_4 in intact cells incorporates [³H]inositol at the same rate as [³H]Ins(1,3,4,5,6)P_5; as for PP-InsP_5 and [PP]$_2$-InsP_4, they incorporate [³H]inositol at the same rate as [³H]InsP_6 (38, 42). Therefore, the study of their turnover *in situ* requires prolonged labelling times with [³H]inositol to ensure labelling to isotopic equilibrium (see section 2). Specific quench, extraction and HPLC procedures have had to be developed to quantify changes in the levels of these compounds in cells (*Protocol 4* and *Figure 4*).

Protocol 4. The HPLC assay of cellular inositol pyrophosphates

Equipment and reagents

- An HPLC system (see *Protocol 3*)
- A 5 μ Whatman 125 × 4.6 mm PartiSphere SAX HPLC column
- 100 ml Quench buffer: 10 mM NaF, 5 mM Na_2EDTA, 1 mg ml^{-1} InsP_6, 0.1% (v/v) Triton X-100

- 100 NENSORB^{TM} deproteinizing columns (NEN-DuPont)
- 1000 ml HPLC Buffer A (1 mM Na_2EDTA)
- 1000 ml HPLC Buffer B (1.3 M $(NH_4)_2HPO_4$ pH 3.8 with H_3PO_4 plus 1 mM Na_2EDTA; dissolve EDTA first)

Method

1. Terminate incubations by aspiration of the incubation medium, and then lyse cells immediately by addition of 1.5 ml ice-cold quench buffer. Keep extracts on ice for 15 min prior to centrifugation to remove cell debris. Save and deproteinize the resultant supernatants (step 3).

2. Prime a NENSORB^{TM} preparative deproteinizing column with 10 ml methanol followed by 10 ml quench buffer. The flow-rate should be kept to approx. 3 $ml\,min^{-1}$ by applying either positive or negative pressure. Monitor the point at which the meniscus of one eluate moves below the frit that protects the resin; this is when the next eluate is added. This precaution is necessary because the columns must not dry out while in use.

3. Add the supernatant from step 1 (i.e. the quenched reaction) to the column, and then wash with 2 × 4 ml aliquots of quench medium. Save and combine these 'flow-throughs'. Samples may be stored at −20 °C for further analysis.

4. Load samples onto a Partisphere 5 μ SAX HPLC column at a flow rate of 1 $ml\,min^{-1}$.

5. Elute (at 1 $ml\,min^{-1}$) according to the following procedure: 0–10 min, B=0%; 10–30 min, B increases linearly to 40%; 30 to 130 min, B increases linearly to 75%; 130–131, B increases linearly to 100%; 131–157 B=100%; 157 min, B returns to 0%. 1 ml fractions are collected. Mix all HPLC fractions with 4 volumes of Monoflow 4 scintillant (National Diagnostics, Manville, NJ) and quantify radioactivity by liquid scintillation spectrometry.

---

4. Analysis of inositol phosphate turnover in cell-free extracts

The metabolism of an [^{3}H]-labelled inositol phosphate by both cell-free extracts and purified enzymes can be assayed by gravity-fed columns (*Protocol 5*).

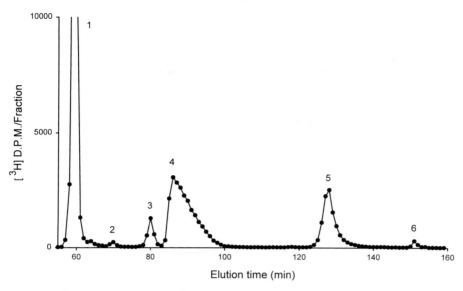

Figure 4. Partisphere SAX analysis of inositol 'pyrophosphates'. DDT₁MF-2 hamster smooth muscle cells were labelled with 20 μCi ml⁻¹ [³H]inositol for 5 days. They were treated with 0.8 mM NaF for 40 min to inhibit inositol 'pyrophosphate' phosphatases and expose the underlying rate of inositol 'pyrophosphate' synthesis (37). Cells were quenched and analysed on a Partisphere SAX column as described in *Protocol 4*. Peak 1 = Ins(1,3,4,5,6)P_5, peak 2 = Ins(1,2,4,5,6)P_5, peak 3 = PP-InsP_4, peak 4 = InsP_6, peak 5 = PP-InsP_5, peak 6 = [PP]₂-InsP_4.

If instead the assays are to be analysed by HPLC (e.g. as described in *Protocol 3*), the incubations should first be quenched and neutralized as described in steps 2 and 5 of *Protocol 1* to remove all traces of protein.

Protocol 5.   Assay of cell-free inositol phosphate metabolism by gravity-fed ion-exchange columns

Equipment and reagents

- 500 g ion-exchange resin, AG 1-X8 200–400 mesh, formate form (Bio-Rad)
- 100 Polyprep propylene column supports (Bio-Rad)
- 100 ml 0.2M ammonium formate/0.1 M formic acid/0.2 mg ml⁻¹ InsP_6
- 1 l 0.4 M ammonium formate/0.1 M formic acid

- 1 l 0.18 M ammonium formate
- 1 l 0.8 M ammonium formate/0.1 M formic acid
- 1 l 1.05 M ammonium formate/0.1 M formic acid
- 1 l 2 M ammonium formate/0.1 M formic acid

Method

1. Set up the required number of assays (50–200 μl volume) using the appropriate buffer, tissue extract and [³H]inositol phosphate.

Protocol 5. *Continued*

2. Quench assays with 1 ml 0.2 M ammonium formate/0.1 M formic acid/0.2 mg ml^{-1} InsP$_6$.

3. Stand tubes on ice for 20 min, then centrifuge at 10 000 *g* for 5 min. Remove supernatant, and apply to an ion-exchange column prepared as described in *Protocol 2*.

4. Elute Ins*P* with 2 × 5 ml aliquots of 0.18 M ammonium formate[a].

5. Elute Ins*P*$_2$ with 2 × 5 ml aliquots of 0.4 M ammonium formate/0.1 M formic acid[a].

6. Elute Ins*P*$_3$ with 2 × 5 ml aliquots of 0.8 M ammonium formate/0.1 M formic acid[a].

7. Elute Ins*P*$_4$ with 2 × 5 ml aliquots of 1.05 M ammonium formate/0.1 M formic acid[a].

8. Elute Ins*P*$_5$ with 1 × 3 ml aliquot of 2 M ammonium formate/0.1 M formic acid[a].

[a]These elution protocols can only be considered to be a guide. The optimal volumes and concentrations of the various eluates will need to be determined empirically for each batch of resin using appropriate standards.

References

1. Berridge, M.J. (1993). *Nature*, **361**, 315–325.
2. Biden, T.J., Prugue, M.L., and Davidson, A.G.M. (1992). *Biochem. J.*, **285**, 541–549.
3. Batty, I.H. and Downes, C.P. (1995). *J. Neurochem.*, **65**, 2279–2289.
4. Shears, S.B. (1991). *Pharmac. Ther.*, **49**, 79–104.
5. Stephens, L.R. and Downes, C.P. (1990). *Biochem. J.*, **265**, 435–452.
6. Tarver, A.P., King, W.G., and Rittenhouse, S.E. (1987). *J. Biol. Chem.*, **262**, 17268–17271.
7. Batty, I.H., Michie, A., Fennel, M., and Downes, C.P. (1993). *Biochem. J.*, **294**, 49–55.
8. Irvine, R.F., Letcher, A.J., Lander, D.J., Drøbak, B.K., Dawson, A.P., and Musgrave, A. (1989). *Plant Physiology*, **89**, 888–892.
9. Coté, G.G., Quarmby, L.M., Satter, R.L., Morse, S.M.J., and Crain, R.C. (1990). In *Inositol metabolism in plants* (ed. D.J. Morré, W.F. Boss, and F.A. Loewus), pp. 113–137. Wiley-Liss, New York.
10. Irvine, R.F., Letcher, A.J., Stephens, L.R., and Musgrave, A. (1992). *Biochem. J.*, **281**, 261–266.
11. Batty, I.H. and Downes, C.P. (1993). *Biochem. J.*, **297**, 529–537.
12. Horstman, D.A, Takemura, H., and Putney, J.W., Jr. (1988). *J. Biol. Chem.*, **263**, 15 297–15 303.
13. Lakin-Thomas, P.L. (1993). *Biochem. J.*, **292**, 805–811.

14. Nahorski, S.R., Ragan, C.I., and Challiss, R.A.J. (1991). *Trends Pharmacol.Sci.*, **12**, 297–301.
15. Menniti, F.S., Oliver, K.G., Nogimori, K., Obie, J.F., Shears, S.B., and Putney, J.W., Jr. (1990). *J. Biol. Chem.*, **265**, 11 167–11 176.
16. Coté, G.C., Yueh, Y.G., and Crain, R.C. (1996). In *Subcellular biochemistry* (ed. B.B. Biswas and S. Biswas), Vol. 26, pp. 317–343. Plenum Press, New York.
17. Menniti, F.S., Takemura, H., Oliver, K.G., and Putney, J.W., Jr. (1992). *Mol. Pharm.*, **40**, 727–733.
18. Wreggett, K.A., Howe, L.R., Moore J.P., and R.F. Irvine (1987). *Biochem. J.*, **245**, 933–936.
19. Brown, J.E., Rudnick, M., Letcher A.J., and R.F. Irvine (1988). *Biochem. J.*, **253**, 703–710.
20. Los, G.V., Artemenko, I.P., and Hokin, L.E. (1995). *Biochem. J.*, **311**, 225–232.
21. Wong, N.S., Barker, C.J., Shears, S.B., Kirk, C.J., and Michel, R.H. (1988). *Biochem. J.*, **252**, 1–5.
22. Hawkins, P.T., Stephens, L.R., and Piggott, J.R. (1993). *J.Biol.Chem.*, **268**, 3374–3383.
23. Rincón, M., Chen, Q., and Boss, W.F. (1988). *Plant Physiol.*, **89**, 126–132.
24. Singh, J., Hunt, P., Eggo, M.C., Sheppard, M.C., Kirk, C.J., and Michell, R.H. (1996). *Biochem. J.*, **316**, 175–182.
25. Mayr, G.W. (1990). In *Methods in inositide research* (ed. R.F. Irvine), pp. 83–108. Raven Press, New York.
26. Li, G., Pralong, W.-F., Pittet, D., Mayr, G.W., Schlegel, W., and Wollheim, C.B. (1992). *J. Biol. Chem.*, **267**, 4349–4356.
27. Bunce, C.M., French, P.J., Allen, P., Mountford, J.C., Moor, B., Greaves, M.F., Michell, R.H., and Brown, G. (1993). *Biochem. J.*, **289**, 667–673.
28. French, P.J., Bunce, C.M., Stephens, L.R., Lord, J.M., McConnell, F.M., Brown, G., Creba, J.A., and Michell, R.H. (1991). *Phil. Trans. R. Soc. Lond.*, B **245**, 193–201.
29. Eisenberg, F., Jr. (1967). *J. Biol. Chem.*, **242**, 1375–1382.
30. Lee, Y., Choi, Y.B., Suh, S., Lee, J., Assmann, S.M., Joe, C.O., Kelleher, J.F., and Crain, R.C. (1996). *Plant Physiol.*, **110**, 987–996.
31. McConnell, F.M., Shears, S.B., Lane P.J.L., and Clark, E. A. (1992). *Biochem J.*, **284**, 447–455.
32. Murphy, P.N.P. (1996). In *Subcellular biochemistry* (ed. B.B. Biswas and S. Biswas), Vol. 26, pp. 317–343. Plenum Press, New York .
33. Barker, C.J., French, P.J., Moore, A.J., Nilsson, T., Berggren, P.-O., Bunce, C.M., Kirk, C.J., and Michell, R.H. (1995). *Biochem. J.*, **306**, 557–564.
34. Kirk, C.J., Morris, A.J., and Shears, S.B. (1990). In *Peptide hormone action: a practical approach* (ed. K. Siddle and J. Hutton), pp. 151–182. IRL Press, Oxford.
35. Stephens, L.R. (1990). In *Methods in inositide research* (ed. R.F. Irvine), pp. 9–30. Raven Press, New York.
36. Cho, M.H., Tan, Z., Erneux, C., Shears, S.B., and Boss, W.F. (1995). *Plant Physiol.*, **107**, 845–856.
37. Shears, S.B., Ali, N., Craxton, A., and Bembenek, M. E. (1995). *J. Biol. Chem.*, **270**, 10 489–10 497.
38. Menniti, F.S., Miller, R., Putney, J.W., Jr., and Shears, S.B. (1993). *J. Biol. Chem.*, **268**, 3850–3856.

39. Stephens, L.R., Radenberg, T., Thiel, U., Vogel, G., Khoo, K.-H., Dell, A., Jackson, T.R., Hawkins, P.T., and Mayr, G.W. (1993). *J. Biol. Chem.*, **268**, 4009–4015.
40. Voglmaier, S.M., Bembenek, M.E., Kaplin, A.I., Dormán, G., Olszewski, J.D., Prestwich, G.D., and Snyder, S.H. (1996). *Proc. Natl. Acad. Sci. U.S.A.*, in press.
41. Laussmann, T., Eujen, R., Weisshuhn, C.M., Thiel, U., and Vogel, G. (1996). *Biochem. J.*, **315**, 715–725.
42. Glennon, M. C. and Shears, S.B. (1993). *Biochem. J.*, **293**, 583–590.
43. Estevez, F., Pulford, D., Stark, M.J.R., Carter, A.N., and Downes, C.P. (1994). *Biochem. J.*, **302**, 709–716.
44. Stephens, L.R. and Irvine, R.F. (1990). *Nature*, **346**, 580–583.
45. Van Dijken, P., de Haas, J.-R., Craxton, A., Erneux, C., Shears, S.B., and van Haastert, P.J.M. (1995). *J.Biol.Chem.*, **270**, 29724–29731.
46. Brearley, C.A. and Hanke, D.E. (1996). *Biochem. J.*, **314**, 227–233.

3

Mass assay of inositol and its use to assay inositol polyphosphates

TAMAS BALLA

Myo-inositol was discovered in 1850 as a component of muscle tissue, but its importance was not recognized until about 100 years later when deprivation of dietary inositol in various animals was shown to cause disorders including alopecia and lipodystrophy in liver and intestinal epithelial cells (see ref. 1 for review). These changes could be reversed by administration of inositol and increased in severity when the intestinal flora was destroyed, since inositol is probably produced by intestinal microorganisms. Furthermore, several mammalian cell lines have been found to require inositol to sustain growth (2), although some cells can grow without added inositol because they can synthesize sufficient amounts to support their needs (3).

The finding that *myo*-inositol is a component of membrane lipids (4), and the subsequent characterization of these lipids as phosphoinositides (5), began a new era that led to the description of the inositide-derived second messenger system (6) and the recognition of its central role in calcium signaling (7). While these developments explained why *myo*-inositol is an essential component for normal cellular functions, more recent observations have indicated that this may not be the only aspect of *myo*-inositol metabolism. First, the discovery of 3-phosphorylated inositol lipids and the complex regulation of PtdIns 3-kinases has created a series of new questions about novel and unresolved functions of membrane inositides (8). Second, there has been a proliferation of inositol phosphate isomers identified in various cells and ample evidence to suggest that they may not originate from the second messenger, inositol 1,4,5-trisphosphate, that is produced from membrane phosphoinositides by the action of phospholipase C. Cells contain large amounts of highly phosphorylated inositols, such as InsP$_5$ and InsP$_6$ and also pyrophosphorylated polyphosphoinositols (9, 10). These compounds are synthesized through a series of phosphorylations from inositol 3-phosphate in *Dictyostelium* (11), but their synthetic route is a matter of debate in mammalian cells (12). The role of these compounds is not clear at present, and they have been proposed to serve as antioxidants, phosphate reserves, neurotransmitters or regulators of membrane fusion events (see refs. 13 and 14). It is interesting that their synthesis is increased during the S-phase of the cell cycle (15), a finding that accounts for their high degree of labeling with *myo*-[^{3}H]inositol in rapidly growing cells. They are also found in high concentrations in germinating seeds, but cells that do not divide (such as avian erythrocytes) can also produce InsP$_5$, although not InsP$_6$ (16).

Progress in the understanding the role of highly phosphorylated inositols in cell regulation has been hampered by the delicate, and often elaborate methodology that is required to perform experiments on inositol polyphosphates, and by the frustration caused by the complexity of their metabolic relationships. Most of our knowledge on phosphoinositides relies on experiment using either [^{32}P]phosphate, or more often, *myo*-[^{3}H]inositol, to metabolically label cells prior to analysis of the labeled lipids and/or inositol phosphates (see Chapters 1 and 2 of this book). While these studies were of great value in reaching our current understanding of the pathways of inositol phosphate metabolism, they have their limitation in that changes in isotope labeling does not always reflect changes in mass, depending on the labeling time and the relative turnover rate of the various metabolic pathways. This is especially true for InsP$_5$ and InsP$_6$ pools since the turnover of some of their phosphate groups is metabolically more sluggish. Mass measurements have been developed for Ins(1,4,5)P$_3$ based on radio receptor assays (17), or on a metal-dye detection system (18) and more recently an ELISA assay for Ins(1,4,5)P$_3$ has been described (19). However, only the metal-dye detection method is available for measurements of other inositol phosphates and it cannot measure *myo*-inositol itself. Therefore, measurement of either the phosphate or *myo*-inositol content of the purified inositol containing species remains one of the few tools to estimate the absolute mass of these compounds. Since phosphate is a component of numerous soluble molecules, that may be difficult to separate from the particular inositol phosphate of interest, assaying the *myo*-inositol content provides a far greater specificity.

This chapter describes a specific and sensitive assay system for the analysis of *myo*-inositol together with its application to measure inositol polyphosphates. This method is largely based on the procedure developed by Maslansky and Busa (20). While their description of this method is quite detailed and extremely valuable, we introduced some changes to improve its reproducibility and to extend its use to measurements of highly phosphorylated inositols. In this chapter, we give a detailed description of each step even where their original protocol was unchanged. The basic reaction of the assay is the conversion of *myo*-inositol into *scyllo*-inosose by the enzyme, *myo*-inositol dehydrogenase (IDH, EC. 1.1.1.18) in the presence of NAD$^+$ (21), and the quantification of the generated NADH by a coupled reaction using the enzyme, diaphorase (EC.1.8.1.4.), and resazurin as the electron acceptor substrate (22). Upon reduction, resazurin is converted to resorufin, which is highly fluorescent and can be readily quantified in a fluorimeter. As simple as

ANALYTICAL STEPS   MATERIALS

I.   Extraction

perchloric acid
tri-n-octylamine
(1,1,2-trichloro-trifluoroethane
phenol red
EDTA

II.   Chromatography

Sep Pak Accel Plus QMA minicolumns
Triethylamine
CO_2
labeled inositol phosphate standards

III.   Dephosphorylation

phytase
TMD-8 mixed ion exchange resin
mannitol

IV.   Assay of *myo*-inositol

myo-inositol (unlabeled)
inositol dehydrogenase (IDH)
NAD+
resazurin
diaphorase

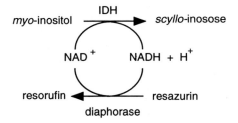

Figure 1. The principle of *myo*-inositol assay and the analytical steps for its application to mass determinations of inositol polyphosphates.

it sounds, one needs to consider a number of technical details to make this method quantitative and specific. The following is a point by point discussion of the individual steps from the extraction through the assay and quantification (*Figure 1*) with special emphasis on technical details.

1. Extraction of inositol phosphates

Tissue preparation and extraction is essentially the same as for the measurement of radio-labeled inositol phosphates and such methods are widely available in the literature (23). Most of these methods are based on the precipitation of the proteins (usually entrapping some of the lipids) by either TCA or PCA, and separating the water soluble products from the precipitate by centrifugation. After removal of TCA (by extraction with diethyl ether) or PCA (with KOH, which forms insoluble potassium perchlorate, or with the mixture of Freon and tri-*n*-octylamine, which reacts with PCA and forms a separate phase) the sample is ready for further analysis. It is important to consider the following points:

(a) Inositol phosphates, especially those with more than three phosphates like to stick to glass surfaces even from solutions. This is particularly true if the sample does not contain other, more abundant highly negatively charged compounds. Therefore, we usually use plastic (polypropylene) tubes for inositol phosphate analysis.

(b) Diethyl ether is very volatile and flammable and should be used in a chemical hood. TCA is therefore replaced by PCA in most extraction procedures. Additionally, any traces of TCA that are not removed can inhibit dehydrogenases.

(c) The easiest method to remove PCA is the use of KOH. The potassium perchlorate that is formed is water-insoluble but this process is temperature and time-dependent and also requires the pH to be monitored (should be >7.0). Also, the precipitate (which cannot be washed again) traps some of the inositol phosphates after centrifugation. Detailed description of this method is found in (20).

(d) The use of a 1:1 mixture of Freon and tri-*n*-octylamine (24) is a rapid and reliable method to extract PCA from the samples, but the chlorofluorocarbon component may limit its future use and should be an environmental concern for large-scale applications. This method can lead to low recoveries of highly phosphorylated inositols if present at low amounts (25). With the amounts of tissues that are required to obtain measurable amounts of InsPs, we did not encounter recovery problems.

Since we usually analyze the inositol phosphate content of cultured cells the following is a description of the protocol used in our laboratory:

Protocol 1. Extraction of inositol phosphates from cultured cells

Reagents

- Perchloric acid (diluted to 15% weight/vol. with deionized water)
- Trioctylamine (Fluka)
- Freon (1,1,2-trichloro-triflouroethane, Fluka)
- 1 M EDTA (use Na^+ and not K^+ salt)
- phenol red (5 mg ml^{-1}, use Na^+ salt, or add NaOH to bring into solution)
- Dulbecco's phosphate buffered saline (PBS, w/o Ca^{2+} and Mg^{2+})

Method

1. Grow cells on 10 cm diameter culture dishes to about 70–80% confluence[a], and treat them according to the experimental protocol.

2. Wash the plates rapidly two times with 3 ml of ice-cold PBS.

3. Add 1 ml each of ice-cold PBS and PCA (15%, w/v).

4. Scrape cells with a cell lifter and transfer to a 15 ml conical tube.

5. Wash the plate with 0.5 ml of distilled water, and combine this wash in the same tube.

6. After two freeze-thaw cycles (this step is important to release inositol phosphates!), let the samples stand on ice for ~20 min before centrifuging at 2000 × *g* at 4°C for 20 min.

7. Withdraw the supernatant and transfer into another 15 ml tube that contains 10 μl 1 M EDTA and 1 μl phenol red

8. Vortex samples briefly and add 2 ml of a 1:1 mixture of Freon and tri-*n*-octylamine (freshly made) to the tubes. Vortex samples very thoroughly for 1 min at room temperature.

9. Centrifuge tubes at room temperature for 5 min at 2000 × *g* and withdraw the top layer[b] with care (because of the surface tension of the oil, it is tricky to remove the last bits of the water phase) and transfer into a separate 15 ml tube.

10. Samples are slightly acidic at this point and should be neutralized just prior to chromatography. Usually 3–4 μl of 1 N NaOH is sufficient[c] and the use of KOH should be avoided as it can cause some precipitation (see above).

[a]Cell density affects numerous signaling events and may influence inositol phosphate levels. It is advised that cells are always grown to the same density for reproducible results.
[b]Three phases of liquid should be visible in the tubes at this point. The upper layer should be colorless, this contains the water-phase with the inositol phosphates. Below this is a yellow oily layer that contains the octylamine-Freon mixture that reacted with PCA, and on the bottom, a small amount of colorless layer with unreacted Freon-octylamine. The use of phenol red helps discriminating between the layers and it is important to have some unreacted reagent in the bottom ensuring that it was present in excess amounts.
[c]The easiest way to monitor the pH during this step is to add 1 μl phenol red to the sample. The yellow color of the extracted sample should turn to orange-red after NaOH addition, but pink color means overalkalinization.

The recovery of [³H]InsP$_6$ added to the samples before extraction has been found to be greater than 90% with this procedure even without re-extraction of the freon/trioctylamine mixture.

2. Separation of inositol phosphates

Obviously, the specificity of the overall assay relies upon the separation of the particular inositol phosphate of interest from the other inositol phosphates present in the sample. It really depends on the scope of the study, how demanding this step should be. To separate the major groups of inositol phosphates (based on increasing charge) from one another is relatively simple up to $InsP_4$, but is slightly more difficult if $InsP_5$ and $InsP_6$ is included in the series. If the isomeric forms within these groups need to be separated, the only choice is HPLC. Numerous such methods have been described in the literature (25–27) and are not discussed further in this chapter. The key problem in the further processing of such HPLC-purified samples is their high salt content that is necessary for their elution from the ion exchange columns. Desalting steps are available, but most of them cause substantial loss of the sample. To overcome this problem several attempts have been made to use salts for elution that can be lyophilized away and does not require desalting of the sample. Important points to consider:

(a) Phosphate buffers hould be avoided as phosphate may interfere with the subsequent dephosphorylation steps.

(b) Ammonium salts should be avoided as they inhibit some dehydrogenases.

These constraints narrow down the available options and make the use of Sep Pak columns the best compromise (20, 28). These columns are fast and simple, and the TEAB (triethylammonium bicarbonate) used for elution can be completely removed by vacuum centrifugation. Stepwise elution of $InsP_1$ $InsP_2$ and $InsP_3$ has been described (20) but separation of higher inositol phosphates is difficult (if not impossible) by stepwise elution (*Figure 2A*). We

---

Figure 2. Separation of inositol phosphates on Sep Pak Accel Plus QMA columns. (A) Stepwise elution with TEAB. *myo*-[^{3}H]inositol labeled inositol phosphate standards (about 15 000 cpm each) were loaded on five separate prewashed Sep Pak columns in cell extracts (obtained from NIH 3T3 cells grown on a 10 cm culture dish, see *Protocol 1* for details), diluted to 4 ml with deionized water. The columns were washed with 4 ml deionized water and the inositol phosphates eluted with 4 ml elution steps of the appropriate concentrations of TEAB solution as shown in the figure. Usually, two 4 ml elution steps at each concentration was sufficient to elute the inositol phosphates up to Ins(1,4,5)P_3, but Ins(1,3,4,5)P_4 required 4×4 ml to be completely eluted. No significant elution of the inositol phosphates was observed at the TEAB concentrations used to elute the preceding, lower inositol phosphate species. Rapid elution of $InsP_5$ and $InsP_6$ together was achieved with elution by 1.5 M TEAB following the 0.5 M washes to remove $InsP_4$ and lower InsPs. (B) Elution by gradient elution with TEAB. *myo*-[^{3}H]inositol labeled inositol phosphate standards (Ins, Ins(1)P, Ins(1,4)P_2, Ins(1,4,5)P_3, Ins(1,3,4)P_3, Ins(1,3,4,5)P_4 and InsP$_6$) were mixed together in tissue extracts (see above), diluted with deionized water and loaded on SepPak columns with an HPLC system. Elution was achieved with a gradient of TEAB (as indicated) with a flow rate of 1 ml min^{-1} and the radioactivity of the effluent was monitored with an on-line radioactive flow detector. The dashed lines show the elution pattern of labeled inositol phosphate standards and the gray area below the solid line shows that of cell extracts obtained from NIH 3T3 cells prelabeled with *myo*-[^{3}H]inositol for 48 h. The figure shows that isomers of InsPs (such as Ins(1,4,5)P_3, Ins(1,3,4)P_3 are not separated by this technique.

have extended the use of Sep Pak columns to separate the higher inositol phosphates, but this really requires the use of gradient elution for which we have used our HPLC system equipped with an on-line radioactive flow-detector to optimize the elution pattern (*Figure 2B*). Although these mini column cartridges can be reused after regeneration, we find that the retention times change significantly after 3–4 runs preventing reliable sample collection unless elution is monitored with radioactive standards. The method used in our laboratory is given in *Protocol 2*.

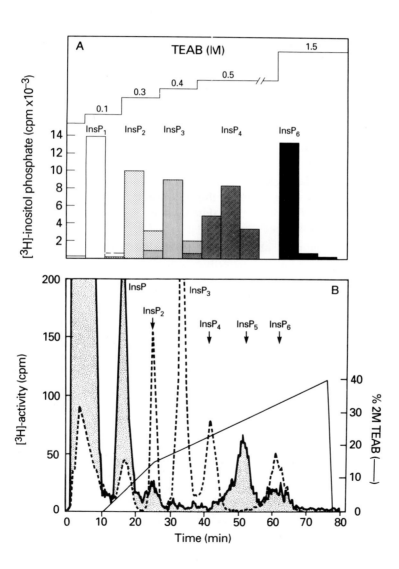

Protocol 2. Separation of inositol phosphates on Sep Pak columns

Equipment and reagents

- Triethylamine bicarbonate (TEAB) solutions should be freshly made at a concentration of 2.0 M (20) (the highest molarity commercially available is 1 M from Fluka)[a]
- Mannitol (10 mM solution in distilled water)
- Sep Pak, Accel plus QMA minicartridges (Waters, Millipore, Milford, MA)

Method

1. Before first use, wash Sep Pak columns with deionized water (10 ml) and then with 2.0 M TEAB (10 ml) at a flow rate of less than 0.5 ml min^{-1}.[b]

2. Wash cartridges thoroughly with deionized water (at least 20 ml) before sample application.

3. Dilute neutralized samples with deionized water to 10–15 ml and apply sample to the column. It is important not to allow the columns run dry after this step. If *myo*-inositol is to be determined the flow-through should be collected with an additional 4 ml wash with deionized water.

4. To elute inositol phosphates with stepwise elution, apply 2–3 successive elutions (4 ml each) with the appropriate concentration of TEAB solution (see *Figure 2A* for details) and collect the fractions into polypropylene tubes.

5. Add 5 µl mannitol at this point to the samples and dry them by lyophilization or vacuum centrifugation.

[a]Triethylamine is chilled on ice and diluted to 2.0 M by slow addition of ice-cold deionized water with constant stirring. This solution is then bubbled with water-saturated CO_2 (passed through deionized water) through a fine sintered glass disc while monitoring the pH (correction is made for temperature) and continued to a pH of 8.4. Acidification continues even after stopping CO_2, therefore it should be approached slowly. It is a good practice to save some of the 2.0 M solution from before CO_2 bubbling, so that the pH can be brought back in case of overacidification. This solution is then filtered through 0.45 µm nylon filters and stored in dark bottles at 4°C. It might be an overprecaution but we never use TEAB solutions stored for longer than 2 days.
[b]The easiest way to drive solutions through the columns is to use a vacuum manifold but we used syringes to apply the solutions during stepwise elutions and an HPLC system (with 1 ml min^{-1} flow rate) for the gradient elutions.

3. Dephosphorylation of inositol phosphates

Unless the *myo*-inositol is analyzed directly, the purified inositol phosphates have to be dephosphorylated before assaying their *myo*-inositol content. Although this step seems to be quite straightforward, it can be a source of problems. While the use of alkaline phosphatase was sufficient to dephosphorylate InsP$_3$, it was found to be far less efficient for Ins(1,3,4,5)P$_4$ (20). In our hands less than 10% of InsP$_5$ and InsP$_6$ was dephosphorylated during treatment with alkaline phosphatase for 3 h at 37°C. Although highly phosphorylated inositols have been successfully dephosphorylated with alkaline phosphatase at pH 7.0 with no Mg^{2+} present (29), we found that treatment with microbial phytase (EC 3.1.3.8.) is a very efficient way to completely dephosphorylate InsP$_5$ and InsP$_6$ in a reasonably short period of time at room temperature (*Figure 3A*).

Protocol 3. Dephosphorylation of inositol phosphates by phytase enzyme

Equipment and reagents

- Phytase (Sigma, P-9792) (8 mg ml^{-1}, 4.1 U mg^{-1} solid, in glycine buffer)
- Poly-Prep chromatography minicolumns (Bio-Rad)
- TMD-8 mixed ion exchange resin (Sigma, M-8157) (batch-washed with deionized water before use)

Method

1. Dissolve dried samples in 0.5 ml distilled water (vortexed thoroughly) and add 0.5 ml of 50 mM glycine buffer (pH 2.5) and 50 μl of phytase to the tubes.

2. After vortexing, incubate samples at room temperature for 3 h.

3. Apply samples to mini columns (Bio-Rad) (with the outlet plugged) containing 0.6 ml prewashed TMD-8 mixed ion exchange resin and vortex several times for about 2 min.

4. Collect samples in Eppendorf tubes and wash the column with 0.5 ml deionized water.

5. After vacuum drying, the samples are ready for the *myo*-inositol assay. Dissolve the sample in 200 μl (or less depending on the expected amount of *myo*-inositol) Tris/ZnCl$_2$/EDTA buffer (see below) and assay several aliquots (10–50 μl).

---

4. Assay to measure *myo*-inositol content

As described above, in the assay *myo*-inositol is oxidized to *scyllo*-inosose by the enzyme, inositol dehydrogenase (IDH) in the presence of NAD$^+$. The produced NADH is then quantitated in a coupled reaction using diaphorase and resazurin (*Figure 1*). These two enzymatic reactions have different pH optima (pH 9.0 for the former and 6.5 for the latter). In the originally described procedure (20) the two reactions were performed in the same tube, but the pH had to be adjusted between the two incubations. However, with

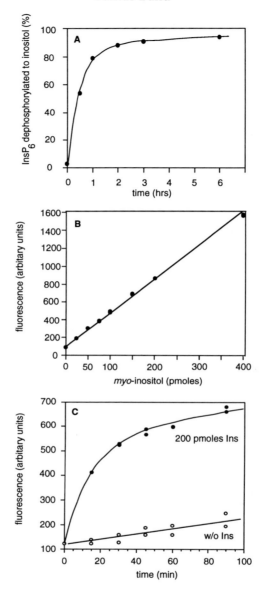

Figure 3. (A) Kinetics of dephosphorylation of $InsP_6$ to *myo*-inositol by microbial phytase. The dephosphorylation of $[^3H]InsP_6$ was monitored as the appearance of $[^3H]$ activity into fractions that did not bind to TMD-8 columns (see *Protocol 3* for details). (B) Standard curve of the *myo*-inositol assay. The *myo*-inositol assay (*Protocol 4*) is linear between 25–400 pmol *myo*-inositol. (C) Kinetics of the assay reaction. The increase in fluorescence as a function of time in the *myo*-inositol assay that contained no added *myo*-inositol (blank, open circles) or 200 pmol *myo*-inositol (closed circles).

this protocol our results were unreproducible and more often than not even a standard curve with *myo*-inositol could not be generated. After checking each individual step in this reaction sequence, we concluded that the variability lies in the extent of NADH reoxidation in the first reaction. Therefore, we tried to find a pH value at which the two reactions could be run simultaneously. This small change made the assay very reliable and reproducible and revealed that the overall pH optimum of the coupled reaction was around pH 8.5, closer to that of the first one. When the components listed below are mixed the pH will be about 8.5. Our method is given in *Protocol 4*.

Protocol 4. Assay for *myo*-inositol content

Equipment and reagents

- Tris/ZnCl$_2$/EDTA buffer (0.1 M/0.1 mM/0.05 mM, pH 9.0)
- NAD$^+$: since NAD$^+$ solutions are relatively labile at alkaline pH, it is preferable to use NAD$^+$ free acid, rather than the sodium salt, to make a fresh 100 mM stock solution each time, which is then kept on ice (add solid crystals of Na-bicarbonate until it dissolves). Just prior to the assay, this solution should be diluted (1:1) with the Tris/ZnCl$_2$/EDTA buffer (pH 9.0) to the desired 50 mM concentration. Also, some batches of NAD$^+$ gave very high blanks, suggesting NADH contamination (0.1% contamination would give 250 pmol NADH$^+$ per tube which is close to the highest amount of inositol in the assay standard curve).
- Diaphorase reagent (Sigma, D-2381, type II-L): This enzyme is provided in lyophilized form and should be reconstituted in 20 mM phosphate buffer (pH 6.8) containing 0.02% BSA at a concentration of 10 U ml^{-1}. It is stored at -70°C in small aliquots and is discarded after the assay.
- Inositol dehydrogenase (IDH) (Sigma, I-5010): This reagent shows great lot-to-lot variation and some lots had no measurable activity in our hand. It is essential to check its activity upon arrival (for assaying its activity follow the manufacturer's instruction). Although the original description of this method (20) advised the use of the highly purified enzyme, we have found that the more crude enzyme (which was much more active) worked as well and did not increase the blank. However, if there is a choice, the purified enzyme is preferred since it is not certain that other lots of the crude enzyme also have low blank values. The enzyme is reconstituted in 10 mM phosphate buffer (pH 6.8) containing 0.02% BSA at 5 U ml^{-1} concentration and stored in small aliquots at -70°C. Once thawed, the enzyme cannot be stored and used again.
- Resazurin (Aldrich or Sigma): Unfortunately, commercially available resazurin always contains small amounts of contaminants, including a small amount of resorufin (the fluorescent product of the second reaction) that must be removed before it can be used. However, resorufin can be easily separated from resazurin by thin layer chromatography (TLC) on Silica Gel 60 plates. Resazurin is dissolved in chloroform : methanol (2:1) at a concentration of about 0.5 mg ml^{-1}. Roughly 0.5 ml of this solution is applied as a line on to a TLC plate that had been previously activated by heating to 80°C for about 1 h. After the resazurin is dried, the plate is placed in a TLC tank that contains 100 ml of chloroform : methanol (4 : 1) and two drops of 1 M KOH. After development and drying (preferably in the dark) the plate will show a strong purple band of resazurin preceded by a pink line of resorufin. The resazurin band is scraped off the plate into a 12 × 75 glass tube, and the dye is eluted from the silica with 3 ml of 1 M phosphate buffer (pH 6.5) with vigorous vortexing followed by centrifugation at 2000 × *g* for 10 min. The elution is repeated and the two supernatants are combined. then resazurin is quantified by spectrophotometry at 600 nm based on its molar extinction coefficient (44640 with 1 cm path length). It is desirable to store this solution at concentrations > 40 μM at 4°C in the dark. Although resazurin it is stable at room light, it is rapidly degraded in direct sunlight. This solution is stable for 1–2 weeks but for maximum sensitivity a reagent blank should be checked before the assay.
- Fluorescent spectrophotometer

Protocol 4. *Continued*

Method

1. Add the following components to 12 × 75 polycarbonate tubes under dimmed light conditions:

   (i)   Sample or standard (known amount of *myo*-inositol) in a total of 50 μl of Tris/ZnCl$_2$/EDTA buffer;

   (ii) 5 μl 50 mM NAD;

   (iii) 5 μl diaphorase reagent;

   (iv) 10 μl 25 μM resazurin.

2. Mix by vortexing.

3. Start reaction by adding 5 μl inositol dehydrogenase reagent.

4. Incubate at 37°C for 60 min in the dark.

5. Terminate reactions by adding 920 μl 0.1 M Tris/HCl (pH 9.0), vortex, then place tubes on ice.

6. Transfer solutions into cuvettes and read fluorescence at ex: 565 nm and em: 585 nm (10 nm slit).

A typical standard curve is shown in *Figure 3B*, and should be linear up to 400 pmol inositol (even though the amount of NAD is only 250 pmol in the assay). This shows the advantage of the two reactions run simultaneously, since NAD$^+$ can be regenerated by the second reaction and reused in the first. The minimum amount of *myo*-inositol that was reliably distinguished from the blank (no inositol) was 15–20 pmol per tube. Reaction times can be shortened if necessary (*Figure 3C*).

5. Accuracy of the InsP$_6$ measurement

The overall performance of the whole procedure was tested in experiments where known amounts of InsP$_6$ (2–8 nmol) were added into Krebs Ringer solutions that contained 0.1% BSA. The InsP$_6$ and InsP$_5$ contents were measured after extraction, HPLC-separation on Sep Pak columns, and dephosphorylation using the *myo*-inositol assay. To follow the recovery of InsP$_6$, a tracer amount (0.1 μCi = 6.7 pmol/tube) of [^{3}H]InsP$_6$ (DuPont, NEN, 15 Ci mmol^{-1}) was added to the tubes before the extraction. After extraction, samples were run on HPLC using the Sep Pak columns (*Figure 2B*) and the fractions corresponding to InsP$_5$ and InsP$_6$ collected separately. After freeze drying and dephosphorylation, the samples were dissolved in 200 μl Tris/ZnCl$_2$/EDTA buffer and aliquots were assayed in the final *myo*-inositol assay; the radioactivity of an aliquot was measured to calculate the recovery of InsP$_6$ in the sample. Based on the *myo*-inositol content of the

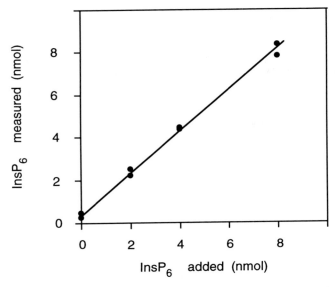

Figure 4. Accuracy of InsP$_6$ mass determination. Various known amounts of InsP$_6$ (0, 2, 4, 8 nmol) were added to PBS containing 0.1% BSA and carried through extraction, HPLC-separation on SepPak columns, dephosphorylation and *myo*-inositol mass assay. The recovery of InsP$_6$ was monitored by [^{3}H]InsP$_6$ added before extraction. The InsP$_6$ content of the original sample was calculated based on the *myo*-inositol assays and the recovery of [^{3}H]activity.

aliquots and the recovery of [^{3}H] activity, the InsP$_6$ content of the original sample was calculated. The result of such an accuracy study is shown in *Figure 4*. The performance of the assay can be expressed by the equation:

$$\text{InsP}_6 \text{ measured} = 0.96 \times [\text{InsP}_6 \text{ added}] + 0.42 \ (r = 0.998)$$

In the InsP$_5$ samples the calculated values of InsP$_6$ were indistinguishable from the blank, showing that no InsP$_6$ 'spilled' into the InsP$_5$ fractions during separation. It has to be noted that although the recoveries of [^{3}H]InsP$_6$ in the individual steps are quite satisfactory, small losses at each stage are inevitable, and the overall loss through the entire procedure was around 50% and the recovery decreased to 10–15% in the blank (where no unlabeled InsP$_6$ was added). The correction for this low recovery and to the fraction of sample assayed is responsible for the difference between the assay blank and that of the accuracy curve. We were not able to significantly improve the recovery of tracer amounts of InsP$_6$ and these values were obtained with 5 nmol mannitol added to the samples after the Sep Pak separation step to decrease non-specific absorption. (Mannitol has been checked and found not to interfere with the *myo*-inositol assay). These findings have to be considered when calculating the overall sensitivity of the procedure. We suggest that the expected amount of material of interest should be at least 1 nmol in the original sample and the recovery be monitored with radioactive standards (see above). The measured $InsP_5$ and $InsP_6$ contents of NIH 3T3 cells were 1.5–2.0 and 0.7–1.2 nmol mg^{-1} protein, respectively (15).

6. Specificity of the assay

Assay specificity is determined basically at two levels: the specificity of the *myo*-inositol assay itself and the separation power of the method that is used to separate the inositol phosphates. The latter question has been addressed (see above) and the specificity of the assay is discussed below. While inositol dehydrogenase is fairly specific for *myo*-inositol, some inositol isomers, such as *scyllo*- and *epi*-inositol, can also serve as substrates for the enzyme (1.6 and 8% of conversion rate, respectively) (30). However, inositol phosphates other than *myo*-inositol based forms have not been observed to our knowledge. There have been reports of the presence of *scyllo*- and *epi*-inositol in rat brain, but at significantly lower levels than *myo*-inositol (31), and together with their low reactivity with IDH they do not pose an important problem.

However, an additional factor that must be considered is the ability of inositol dehydrogenase to utilize glucose, albeit at a much lower rate than *myo*-inositol. For this reason, the glucose content of the sample should be minimal for reliable results. This is not a problem when inositol polyphosphates are measured, since their preparation procedures will separate them from glucose. However, when *myo*-inositol levels are to be measured, hexokinase and ATP can be added to the samples before the assay to convert glucose to its phosphorylated form that is not recognized by IDH. *Protocol 5* is used for this purpose.

Protocol 5. Removal of glucose from samples before *myo*-inositol assay

Reagents

- Hexokinase reagent (50 mM Tris/HCl pH 9.0, 10 mM MgCl$_2$, 100 μM, ATP and   200 U ml^{-1} hexokinase (Sigma)

Method

1. Add 5 μl hexokinase reagent to the 200 μl sample[a] (see *Protocol 3*, step 5) and incubate for 1 h at 37°C.

2. Boil samples for 2 min and assay as described above.

[a]This procedure changes the slope of the standard curve of the subsequent assay, therefore it should also be performed on the blank and the *myo*-inositol standards.

Summary

We have described a method for the measurement of *myo*-inositol and its phosphorylated derivatives. Although this assay requires care and precision, it is reliable and reproducible, and permits the measurement of the absolute mass of inositol phosphates that is not easily achievable with alternative methods (18, 32). Together with other protocols using radioactive isotopes, this procedure should aid further studies on the phosphoinositide signaling system, and help to better understand the diverse roles that these molecules play in cell regulation.

Acknowledgements

The skillful technical assistance of Albert J. Baukal during these studies is greatly appreciated.

References

1. Holub, B.J. (1982). In *Advances in nutritional research* (ed. H.H. Draper), p. 107. Plenum, New York.
2. Eagle, H., Oyama, V.I., Levy, M., and Freeman, A.E. (1957). *J. Biol. Chem.*, **248**, 191.
3. Jackson, M.J. and Shin, S. (1982). *Cold Spring Harbor Conf. Cell Prolif.*, **9A**, 75.
4. Anderson, R.J. and Roberts, E.G. (1930). *J. Biol. Chem.*, **89**, 611.
5. Folch, J. and Wooley, D.W. (1942). *J. Biol. Chem.*, **142**, 963.
6. Michell, R.H. (1975). *Biochim. Biophys. Acta*, **415**, 81.
7. Berridge, M.J. and Irvine, R.F. (1984). *Nature*, **312**, 315.
8. Kapeller, R. and Cantley, L.C. (1994). *BioEssays*, **16**, 565.
9. Balla, T., Baukal, A.J., Hunyady, L., and Catt, K.J. (1989). *J. Biol. Chem.*, **264**, 13 605.
10. Menniti, F.S., Oliver, K.G., Putney, J.W., Jr., and Shears, S.B. (1993). *Trends Biochem. Sci.*, **18**, 53.
11. Stephens, L.R. and Irvine, R.F. (1990), *Nature*, **346**, 580.
12. Dijken, P.V., deHaas, J.-R., Craxton, A., Erneux, C. Shears, S.B., and Van Haastert, P.J. (1995). *J. Biol. Chem.*, **270**, 29 724.
13. Hawkins, P.T., Poyner, D.R., Jackson, T.R., Letcher, A.J., Lander, D.A., and Irvine, R.F. (1993). *Biochem. J.*, **284**, 929.
14. Ali, N., Duden, R., Bembenek, M.E., and Shears, S.B. (1995). *Biochem. J.*, **310**, 279.
15. Balla, T., Sim, S.S., Baukal, A.J., Rhee, S.G., and Catt, K.J. (1994). *Mol. Biol. Cell.*, **5**, 17.
16. Stephens, L.R., Hawkins, P.T., Carter, N., Chahwala, S.B., Morris, A.J., Whetton, A.D., and Downes, C.P. (1988). *Biochem. J.*, **249**, 271.
17. Palmer, S. and Wakelam, M.J.O. (1990). In *Methods in phosphoinositide research* (ed. R.F. Irvine), p. 127. Raven Press, New York.

18. Mayr, G.W. (1990). In *Methods in phosphoinositide research* (ed. R.F. Irvine), p. 83. Raven Press, New York.
19. Shieh, W.-R. and Chen, C.-S. (1995). *Biochem. J.*, **311**, 1009.
20. Maslansky, J.A. and Busa, W.B. (1990). In *Methods in phosphoinositide research* (ed. R.F. Irvine), p. 113. Raven Press, New York.
21. MacGregor, L.C. and Matschinsky, F.M. (1984). *Anal. Biochem.*, **141**, 382.
22. Guilbault, G.G. and Kramer, D.N. (1965). *Anal. Chem.*, **37**, 1219.
23. Wregett, K.A. Howe, L.R., Moore, J.P., and Irvine, R.F. (1987). *Biochem. J.*, **245**, 933.
24. Sharpes, E.S. and McCarl, R.L. (1982). *Anal. Biochem.*, **124**, 421.
25. Stephens, L.R. (1990). In *Methods in phosphoinositide research* (ed. R.F. Irvine), p. 9. Raven Press, New York.
26. Binder, H., Weber, P.C., and Siess, W. (1985). *Anal. Biochem.*, **148**, 220.
27. Patthy, M., Balla, T., and Aranyi, P. (1990). *J. Chromatography*, **523**, 201.
28. Wregett, K.A. and Irvine, R.F. (1987). *Biochem. J.*, **245**, 655.
29. Menniti, F.S., Miller, R.N., Putney, J.W., Jr., and Shears, S.B. (1993). *J. Biol. Chem.*, **268**, 3850.
30. Guderman, T.W. and Cooper, T.G. (1986). *Anal. Biochem.*, **158,** 59.
31. Sherman, W.R., Hipps, P.P., Mauck, L.A., and Rasheed, A. (1978). In *Cyclitols and phosphoinositides* (ed. W.W. Wells and F. Eisenberg, Jr.), p. 279. Academic Press, New York.
32. Rittenhouse, S.E. and King, W.G. (1990). In *Methods in phosphoinositide research* (ed. R.F. Irvine), p. 109. Raven Press, New York.

4

Expression, purification and interfacial kinetic analysis of phospholipase C-γ1

GWENITH A. JONES and DEBRA A. HORSTMAN

1. Introduction

Phospholipase C-γ1 (PLC-γ1) is an important signalling molecule. Activation of PLC-γ1 by receptor tyrosine kinases allows growth factors, such as EGF, to modulate the level of the second messengers inositol 1,4,5-trisphosphate $(\text{Ins}(1,4,5)P_3)$ and diacylglycerol (1–9). In response to EGF, PLC-γ1 rapidly becomes tyrosine phosphorylated and translocates to the membrane (1, 2, 7, 10–17). Very little information is available concerning the structure/function relationships of PLC-γ1; an exception is the finding that tyrosine phosphorylation of PLC-γ1 at Y783 increases PLC-γ1 PtdIns(4,5)P_2 hydrolysing activity in response to growth factors (3). Studies aimed at elucidating the structure/function relationships of PLC-γ1 have been impaired by the lack of expression systems that yield high levels of protein capable of exhibiting high specific activity to PtdIns(4,5)P_2. Another problem has been the use of many different undefined assay systems, which has sometimes resulted in a lack of consensus as to how structural changes modulate PLC-γ1 function. In this article, the description of an baculovirus expression system for PLC-γ1 is presented (18). This expression system allows high expression levels of a PLC-γ1 fusion protein that exhibits high specific activity toward PtdIns(4,5)P_2. A detergent:PtdIns(4,5)P_2 mixed micelle assay system will also be described (19–21), in which PLC-γ1 displays kinetic behavior similar to that described for other lipid-metabolizing enzymes. The use of the baculovirus expression system and the detergent:PtdIns(4,5)P_2 mixed micelle assay system will allow the assessment of various structural changes on PLC-γ1 function.

2. Expression and purification of PLC-γ1

Performing structure/function studies on PLC-γ1 was in the past hampered by the absence of good protein expression systems. Previous structure/function studies of PLC-γ1 were performed using enzyme immunoprecipitated from A-431 cells (22, 23) or purified from bovine brain (24–27). Both sources of enzyme have inherent problems. Purification of PLC-γ1 from bovine brains is long and arduous (16 bovine brains yield 1 mg purified protein after three weeks of purification (25)). Immunoprecipitation is a quick and relatively inexpensive method to specifically isolate PLC-γ1, but doing structure/function studies on physically constrained enzyme may yield artificial results. The problem of identifying an easily obtainable source of PLC-γ1 is overcome with the development of a PLC-γ1 expression system. The use of the baculovirus expression system enables the expression and purification of large quantities of fully functional PLC-γ1 (18).

2.1 Expression of PLC-γ1 in insect cells

To facilitate purification of the expressed PLC-γ1, PLC-γ1 DNA is fused in frame to hexahistidine ((His)$_6$) DNA. The fusion protein is constructed by inserting a 4.2 kb *Bam*H1–*Hind*III fragment of rat PLC-γ1 DNA in frame with (His)$_6$ in a baculovirus transfer vector pBlueBacHis (Invitrogen). This transfer plasmid and linearized AcNPV viral DNA are co-transfected in *Spodoptera frugiperda* (Sf9) cells by lipofection (Invitrogen) according to the manufacturer's instructions. The transfer vector pBlueBacHis contains β-galactosidase gene to allow the identification of recombinant viral plaques by color screening. A high titre viral stock is generated according to manufacturer's instructions. The manual Invitrogen provides with the Bac-N-Blue (Invitrogen) transfection kit contains some of the clearest and easiest to understand procedures available for expression using baculovirus. *Trichoplusia ni* (High Five) (Invitrogen) insect cells plated 5×10^6 cells/100 mm dish are transfected at a multiplicity of infection (MOI) of 10 with baculovirus containing PLC-γ1 DNA (see *Protocol 1*). High Five insect cells are used because recombinant PLC-γ1 expression levels of 8–10% can be obtained compared to an 1% expression level of the recombinant PLC-γ1 fusion protein in Sf9 cells (18). Cells are incubated at 27°C for two days. Maximal expression of recombinant PLC-γ1 levels occurs 48 h after infection. As a control, uninfected High Five cells are cultured for 48 h.

Protocol 1.   Expression of (His)$_6$-PLC-γ1 in high five cells

Equipment and reagents

- High Five cells (Invitrogen)
- High titre PLC-γ1 viral stock (1 × 10^8 p.f.u. ml^{-1})^a
- 27°C incubator
- ExCell medium (J.R.H. Bioscience)
- 100 mm tissue culture dishes
- 5 ml sterile falcon tubes

Method

1. Plate (5) 100 mm dishes with High Five cells (5×10^6 cells/dish).

2. Let cells sit for 30 min until all cells have attached.

3. Aspirate the medium.

4. Dilute 0.5 ml of high titre PLC-γ1 viral stock in 4.5 ml Excell (J.R.H. Bioscience) medium to give a MOI of 10.[b]

5. Incubate the 5 mls of the diluted PLC-γ1 high titre virus with cells for 1 h.

6. Remove the diluted high titre virus.

7. Feed cells with 10 mls ExCell medium (J.R.H. Bioscience).

8. Incubate transfected cells at 27°C for 48 h.

[a] The PLC-γ1 high titre viral stock is prepared according to procedures recommended by Invitrogen.
[b] Multiplicity of infection (MOI)=(titre of virus)(vol of viral stock)/number of insect cells

2.2 Purification of (His)$_6$-PLC-γ1

PLC-γ1 is purified essentially as described by Horstman *et al.* (18), (see *Protocol 2*), except that a low salt (50 mM NaCl) imidazole buffer gradient is used to elute PLC-γ1 from the affinity column and the imidazole buffers do not contain any detergent. A low salt elution buffer is used to facilitate the further purification of PLC-γ1 on an anion exchange column (see *Protocol 3*). The detergent is excluded since the presence of detergents interfere with the detergent:PtdIns(4,5)P_2 mixed micelle assay system. Some proteins can bind to the Ni^{2+}-NTA agarose through non-specific interactions. A high salt buffer will disrupt most non-specific interactions to Ni^{2+}-NTA agarose, therefore after the column is washed with the lysis buffer an additional wash with a high salt buffer is performed.

Proteins from 100 μl of each column fraction are separated on SDS-PAGE and the gel is stained with coomassie blue to detect proteins (*Figure 1*). The (His)$_6$ tagged PLC-γ1 allows a one step purification of PLC-γ1, giving a > 95% purity. Further purification of (His)$_6$-PLC-γ1 is possible using an anion exchange column (*Protocol 3*). Fractions containing (His)$_6$-PLC-γ1 are pooled and the protein concentration determined by BCA assay (Pierce). Typically, 1 mg of (His)$_6$-PLC-γ1 is recovered from 10–12 mg of clarified lysate, giving an effective (His)$_6$-PLC-γ1 expression level of 8–10%. The procedure can easily be scaled up for production of mg quantities of protein. A 50% glycerol solution of the protein is made, and then aliquoted and stored at −80°C. (His)$_6$-PLC-γ1 has been stored in this buffer at −80°C for 3 months with no loss in activity.

Protocol 2. Purification of $(His)_6$-PLC-γ1 on Ni^{2+}-NTA agarose

Equipment and reagents

- Lysis buffer (20 mM HEPES, pH 8.0, 50 mM NaCl, 1% octyl glucoside)
- High Five cells 48 h post infection (see *Protocol 1*)
- Probe sonciator (Branson sonifer 450)
- Ni^{2+}-NTA agarose (Qiagen)
- Glycerol
- Ultracentrifuge
- High salt low imidazole buffer (20 mM HEPES, pH 8.0, 800 mM NaCl, 10 mM imidazole)
- Low salt low imidazole buffer (20 mM HEPES, pH 8.0, 50 mM NaCl, 10 mM imidazole)
- Low salt high imidazole buffer (20 mM HEPES, pH 8.0, 50 mM NaCl, 150 mM imidazole)
- Peristalic pump
- UV monitor
- Fraction collector

Method

1. Scrape the High Five cells into lysis buffer.

2. Lyse the cells by probe sonication for 30 sec at the lowest setting.[a]

3. Clarify the cell lysates by ultracentrifugation at $441\,000 \times g$ for 20 min.

4. Incubate 12–15 mg of cell lysate with 1 ml Ni^{2+}-NTA agarose for > 2 h or overnight at 4°C on a rocker to insure complete mixing.

5. Pack a 1.5×5 cm column with Ni^{2+}-NTA agarose slurry, wash and elute $(His)_6$-PLC-γ1 from column at a flow rate of 0.5 ml min^{-1}. Collect 1 ml fractions.

   (a) Collect the column flow through.

   (b) Wash the column with 10 column volumes of the lysis buffer.

   (c) Wash the column with 10 column volumes of high salt low imidazole buffer.

   (d) Wash the column with 5 column volumes of low salt low imidazole buffer.

   (e) Wash the column with 30 column volumns of a linear imidazole gradient (10–150 mM imidazole).

6. Analyse $(His)_6$-PLC-γ1 purity by SDS-PAGE and staining gel with commassie blue.

7. Pool fractions containing $(His)_6$-PLC-γ1.

8. Make $(His)_6$-PLC-γ1 protein solution 50% glycerol, aliquot and store at −80°C.

[a] Cell lysis should be monitored by microscopy.

2.3 Common Problems Encountered During Purification

One of the drawbacks of this purification scheme is that (His)$_6$-PLC-γ1 elutes off of the Ni^{2+}- NTA agarose column as a broad peak (*Figure 1*). For applications requiring a concentrated protein solution, this necessates a concentration step. Concentration of purified (His)$_6$-PLC-γ1 has proved to be a difficult endeavor. Significant protein loss occurs during most commonly used concentration steps (centrifugal concentration, dialysis). Presumably the enzyme is binding to the surface. Aggregation of the enzyme upon long-term storage is also a problem. The addition of 1 mM DTT and 50% glycerol to the enzyme storage buffer has been effective in preventing aggregation. DTT cannot be added to the buffers used to purify (His)$_6$-PLC-γ1 on Ni^{2+}-NTA agarose column, because DTT will chelate the Ni^{2+}-NTA from the agarose. Both of these problems can be prevented by the addition of detergents to all buffers. Unfortunately, the presence of detergents will interfere with PLC activity assays. If detergents will not interfere with subsequent applications, it is recommended that 1% octyl glucoside be added to all buffers.

(His)$_6$-PLC-γ1 elutes from the Ni^{2+}-NTA agarose column in 75–80 mM

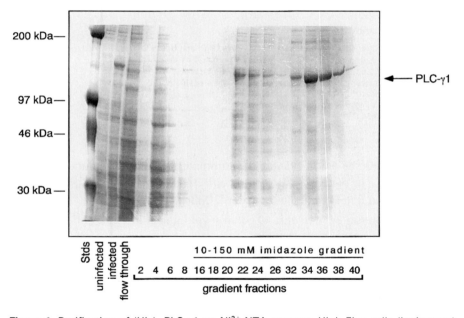

Figure 1. Purification of (His)$_6$-PLC-γ1 on Ni^{2+}-NTA agarose. High Five cells (Invitrogen) were infected with (His)$_6$-PLC-γ1 containing DNA at an MOI of 10 (see *Protocol 1*). (His)$_6$-PLC-γ1 was prepared for purification as described in *Protocol 2*. (His)$_6$-PLC-γ1 was eluted from the Ni^{2+}-NTA agarose column with a 10–150 mM imidazole gradient. Instrumentation used for the purification was a BioRad Automated Econo System run at a flow rate of 0.5 ml min^{-1}, collecting 1 ml fractions. 100 µl of each column fraction was run on a SDS-PAGE gel. The gel was stained with commassie blue to detect protein.

imidazole. Imidazole inhibits PLC-γ1 activity. (His)$_6$-PLC-γ1 assayed in the presence of 0.64 mM imidazole had a specific activity of 0.26 μmol min^{-1} mg^{-1}. (His)$_6$-PLC-γ1 assayed in the absence of imidazole had a specific activity of 2.25 μmol min^{-1} mg^{-1}, an 8.8-fold increase in activity. Therefore, removal of the imidazole is desirable. As mentioned previously, commonly used desalting techniques result in significant loss of protein, and therefore the use of another technique is desirable. Success has been achieved both by changing the buffer and by some concentration of (His)$_6$-PLC-γ1 by further purification of the enzyme over an anion exchange column (*Protocol 3*). (His)$_6$-PLC-γ1 eluted from the Ni^{2+}-NTA agarose column can be directly applied to an anion exchange column. (His)$_6$-PLC-γ1 eluted from the anion exchange column is in a buffer that is compatible with most subsequent applications. The pooled (His)$_6$-PLC-γ1 fractions can now be made 50% glycerol, aliquoted and stored at $-80\,^{\circ}$C.

Protocol 3. Purification of (His)$_6$-PLC-γ1 on anion exchange column

Equipment and reagents

- High Q cartridge, 1 ml (BioRad)
- Equilibration buffer (20 mM HEPES, pH 8.0, 50 mM NaCl)
- (His)$_6$-PLC-γ1 from Ni^{2+}-NTA agarose column (see *Protocol 2*)
- Low salt buffer (20 mM HEPES, pH 8.0, 50 mM NaCl, 2 mM DTT)
- High salt buffer (20 mM HEPES, pH 8.0, 1 M NaCl, 2 mM DTT)
- Glycerol
- Peristaltic pump
- UV monitor
- Fraction collector

Method

1. Equilibrate the anion exchange column with equilibration buffer.

2. Apply (His)$_6$-PLC-γ1 eluted from Ni^{2+}-NTA agarose column to the column at a flow rate of 1.0 ml min^{-1}. Collect 1 ml fractions.

3. Wash column with at least 10 column volumes of low salt buffer.[a]

4. Elute (His)$_6$-PLC-γ1 using a linear salt gradient of 0.05–1 M NaCl in 20 mM HEPES, pH 8.0, 2 mM DTT buffer.[b]

5. Analyse (His)$_6$-PLC-γ1 purity by SDS-PAGE and staining gel with coomassie blue.

6. Pool fractions containing (His)$_6$-PLC-γ1.

7. Make (His)$_6$-PLC-γ1 solution 50% glycerol, aliquot and store at $-80\,^{\circ}$C.

[a]Imidazole absorbs strongly at A$_{280}$, therefore monitor A$_{280}$ readings to assure that all imidazole is washed from the column.
[b]Monitor protein elution by observing A$_{280}$.

3. PLC-γ1 assay

Kinetic analysis is a powerful tool when building a model of enzyme structure/
function relationships. By determining the behavior of the enzyme with its
substrate it is possible to assess the impact of structural changes such as tyrosine phosphorylation or deletion of putative domains on the behavior of the
enzyme. Kinetic analysis also allows the determination of how the substrate
environment affects enzyme behavior. The beauty of kinetic analysis is that it
is a technique which yields data rapidly. As mentioned previously, the use of
many different undefined assay systems has sometimes resulted in a lack of
consensus as to how structural changes modulate PLC-γ1 function. It is
important to select a defined assay system in which the enzyme displays
kinetic behavior that can be predicted by a mathematical model. Such a
model exists for lipid-metabolizing enzymes such as PLC-γ1: the interfacial
kinetic model (19–21).

3.1 Interfacial kinetics

Performing kinetic analysis on phospholipases poses unique problems. PLC-
γ1 is a soluble protein, but its substrate, a phospholipid, is hydrophobic.
Therefore the reaction takes place at the lipid:water interface and an inter-

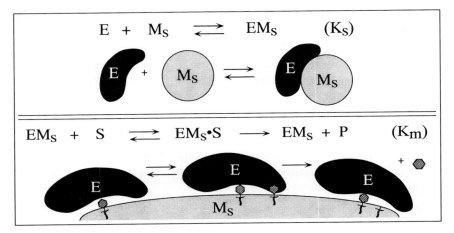

Figure 2. Schematic of the interfacial kinetic model. Interaction of the enzyme with the
water:lipid interface of the detergent:phospholipid mixed micelle or lipid bilayer is
described by the kinetic parameter K_s. This parameter is dependent on both the enzyme
concentration and the molar substrate concentration. Catalysis of substrate at the
micelle or lipid bilayer surface is described by the kinetic parameter K_m. This interfacial
kinetic constant is dependent on the mole fraction of substrate in the micelle or lipid
bilayer. The absolute rate, V_{max}, occurs at infinite molar substrate concentration and infinite
mole fraction.

facial kinetic model is required. The surface dilution model of enzyme kinetics used to describe the kinetic behavior of phospholipase A_2 by Dennis and coworkers (19–21) takes into account reactions that occur at a water:lipid interface (see *Figure 2*). Interaction of the enzyme with the water:lipid interface of the detergent:phospholipid mixed micelle or lipid bilayer is described by the kinetic parameter K_s. This parameter is dependent on both the enzyme concentration and the molar substrate concentration. Catalysis of substrate at the micelle or lipid bilayer surface is described by the kinetic parameter K_m. This interfacial kinetic constant is dependent on the mole fraction of substrate in the micelle or lipid bilayer. The absolute rate, V_{max}, occurs at infinite molar substrate concentration and infinite mole fraction. The surface dilution model has been successfully used to describe the behavior of a number of enzymes that catalyse reactions at a lipid surface (21, 28–31).

The interfacial kinetic behavior of lipid-metabolizing enzymes leads to the proposal of the dual phospholipid model. In this model, the enzyme first binds to the substrate vesicle at a non-catalytic site. Once bound, the enzyme binds substrate at a second lipid binding site, the catalytic site, allowing hydrolysis of substrate to take place (*Figure 2*). PLC-γ1 has been shown to display similar interfacial kinetics to other lipid-metabolizing proteins (22, 23). The kinetic analysis of PLC-γ1 reveals that the non-tyrosine phosphorylated enzyme displays sigmoidal kinetics with respect to substrate (22, 23). Using the Hill equation, the non-tyrosine phosphorylated form of PLC-γ1 has an $S_{0.5}$ of 0.28 mole fraction PtdIns(4,5)P_2, a V_{max} of 0.012 μmol min^{-1} mg^{-1}, an apparent binding constant of 1.2 mM PtdIns(4,5)P_2 and a cooperativity index of 2.5. The tyrosine phosphorylated form of PLC-γ1 does not display sigmoidal kinetics (22, 23), and its kinetic behavior best fits the Michaelis-Menten model of enzyme kinetics. Tyrosine phosphorylation of PLC-γ1 does not effect V_{max}, nor does it significantly effect $S_{0.5}$ (0.2 mole fraction) but it does decrease the apparent binding constant to 0.2 mM PtdIns(4,5)P_2. The major effect of tyrosine phosphorylation on PLC-γ1 is to decrease the cooperativity between the lipid binding sites and to increase the affinity PLC-γ1 has for the substrate micelles. These studies were done using PLC-γ1 immunoprecipitated from A-431 cells (22, 23). Kinetic analysis of (His)$_6$-PLC-γ1 will determine whether it displays kinetic behavior similar to that observed with immunoisolated PLC-γ1.

3.2 PLC assay

The detergent:phospholipid mixed micelle assay system is a quick and efficient means of measuring PLC-γ1 activity. Preparation of the detergent:phospholipid micelles is rapid and requires no special equipment (see *Protocol 4*). Various detergents (Triton X-100, octyl glucoside, sodium deoxycholate and Brij 35) were surveyed as diluents for PtdIns(4,5)P_2 (data not shown). Except for Triton X-100, PLC-γ1 activity in all of the detergents that were surveyed exhibits inconsistent behavior. PLC-γ1 activity could not be correlated with the detergent-critical micelle concentration. It is only in Triton X-100 that PLC-γ1 follows the surface dilution model of enzyme kinetics (22) (data not shown).

Only reagents of the highest quality should be used. Consistent results have been obtained when $[^3H]PtdIns(4,5)P_2$ from Dupont-NEN, $PtdIns(4,5)P_2$ from Boehringer-Mannheim and purified Triton X-100 from Pierce have been used. Care must be taken in the handling of lipids. Any unused portion of $[^3H]PtdIns(4,5)P_2$ should be stored under N_2 or argon at $-80°C$. $PtdIns(4,5)P_2$ from Boehringer-Mannheim comes in 1 mg vials as a lyophilized powder and is stable for at least 12 months at $-20°C$. Typically, the entire 1 mg vial is used to prepare the detergent:phospholipid mixed micelles. The prepared detergent:phospholipid mixed micelles are stable for 1 week at $4°C$.

Protocol 4. Preparation of substrate

Equipment and reagents

- $[^3H]Ptd(4,5)P_2$ (Dupont-NEN)
- $Ptd(4,5)P_2$ (Boehringer-Mannheim)
- Triton X-100 (Pierce)
- Probe sonciator (Branson sonifer 450)

- Phosphate buffer (50 mM phosphate buffer, pH 6.8, 100 mM KCl)
- Speedvac (Savant Instruments)

Method

1. Add 2 μCi of $[^3H]PtdIns(4,5)P_2$ to 1 mg of cold $PtdIns(4,5)P_2$ (Boehringer-Mannheim).

2. Reduce the volume to dryness in the Speedvac (Savant Instruments).

3. Add sufficient phosphate buffer to give a final $PtdIns(4,5)P_2$ concentration of 5 mM.

4. Sonicate the phospholipid suspension with a probed sonicator at the lowest setting on ice until clear.

5. Add Triton X-100 to achieve the desired substrate concentration and desired mole fraction.[a]

[a] Add Triton X-100 to achieve the desired substrate concentration and desired mole fraction according to the following equation:

$$\text{mole fraction } PtdIns(4,5)P_2 = [PtdIns(4,5)P_2]/[PtdIns(4,5)P_2] + [\text{Triton X-100}]$$

Protocol 5. PLC assay

Equipment and reagents

- Purified (His)$_6$-PLC-γ1 (see *Protocols 2 and 3*)
- [^{3}H]Ptd(4,5)P_2 at the desired substrate concentration and mole fraction (see *Protocol 4*)
- Reaction buffer (35 mM NaH$_2$PO$_4$, pH 6.8, 70 mM KCl, 0.8 mM EGTA, 0.8 mM CaCl$_2$)

- 10% cold trichloroacetic acid
- Scintillation fluid
- Scintillation counter
- Microcentrifuge
- 37°C water bath

Method

1. Measure hydrolysis of PtdIns(4,5)P_2 in a reaction volume of 50 μl containing reaction buffer and [^{3}H]PtdIns(4,5)P_2 at the desired substrate concentration and mole fraction.[a] Add 20 ng purified (His)$_6$-PLC-γ1 to each tube.

2. Incubate the assay tubes from 5–20 min at 37°C.[b]

3. Stop the reaction by transferring the tubes to an ice bath.

4. Add 25 μl of ice cold 10% (w/v) trichloroacetic acid and 100 μl of 1% (w/v) bovine serum albumin to each tube.

5. Remove precipitates containing bound [^{3}H]PtdIns(4,5)P_2 from the soluble [^{3}H]Ins(1,4,5)P_3 by centrifugation (13 600 × *g* for 5 min).

6. Transfer all of the supernatant to a scintillation vial.

7. Quantitate the amount of [^{3}H]Ins(1,4,5)P_3 released in the supernatants by scintillation counting.

8. Count an aliquot (10 μl) of the reaction buffer to determine PtdIns(4,5)P_2 specific activity.

9. Calculate the amount of [^{3}H]Ins(1,4,5)P_3 produced.[c]

[a] Include a no enzyme control to determine the amount of background radioactivity.
[b] The incubation time is varied to insure that < 5% of the substrate is hydrolysed.
[c] [^{3}H]Ins(1,4,5)P_3 produced is calculated based on the specific activity of PtdIns(4,5)P_2 (1.7–2.0 cpm pmol^{-1}) using the following equation:

$$\text{pmol } [^3H]\text{Ins}(1,4,5)P_3 = \text{cpm}_{(experimental)} - \text{cpm}_{(no\ enzyme\ control)}/[^3H]\text{PtdIns}(4,5)P_2 \text{ (cpm pmol}^{-1})$$

Activity of (His)$_6$-PLC-γ1 isolated from the baculovirus expression system is linear for 10–40 ng of protein. Typically 20 ng of (His)$_6$-PLC-γ1 is assayed for activity (see *Protocol 5*). The specific activity of the (His)$_6$-PLC-γ1 is comparable to that of PLC-γ1 isolated from bovine brain (2.25 μmol min^{-1} mg^{-1} versus 1 μmol min^{-1} mg^{-1}) (25). From this it is concluded that the (His)$_6$ tag does not interfere with the assay. It is important that the assay conditions are adjusted so that less than 5% of the substrate is hydrolysed. At longer incubation times or at high substrate concentrations (His)$_6$-PLC-γ1 exhibits end-product inhibition (*see Figure 3*). The presence of 10% diacylglycerol in the detergent:PtdIns(4,5)P_2 mixed micelles results in 85% inhibition of (His)$_6$-PLC-γ1 activity (*Figure 4*). For single point determinations, incubating 20 ng of purified (His)$_6$-PLC-γ1 for 10 min at 37°C in the presence of 100 mM, 0.1 mole fraction PtdIns(4,5)P_2, is optimal.

To compare the kinetic parameters of immunoprecipitated PLC-γ1 with

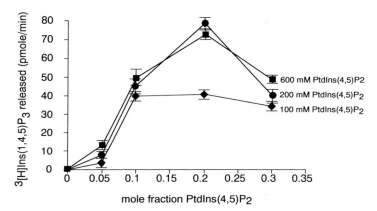

Figure 3. End-product inhibition of (His)$_6$-PLC-γ1. (His)$_6$-PLC-γ1 was assayed at increasing mole fractions of PtdIns(4,5)P_2 (0.05, 0.1, 0.2, 0.3) at 3 molar concentrations of PtdIns(4,5)P_2 (100, 200, 600 mM) using the detergent:PtdIns(4,5)P_2 mixed micelle assay system (see *Protocols 4 and 5*). Each point represents a single determination. Experiments were repeated three times with similar results.

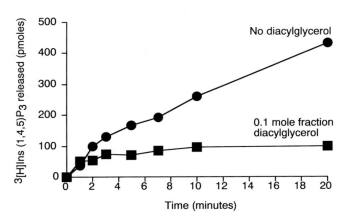

Figure 4. Inhibition of (His)$_6$-PLC-γ1 activity by diacylglycerol. A time course of [^{3}H]IP$_3$ release was performed in the presence or absence of 0.1 mole fraction diacylglycerol. The PtdIns(4,5)P_2 molar concentration in each tube was 100 mM and the micellar concentration of PtdIns(4,5)P_2 was 0.1 mole fraction. Each point represents the average of duplicate measurements.

those of (His)$_6$-PLC-γ1, the enzyme was assayed at three different PtdIns(4,5)P_2 molar concentrations (100, 200, 600 mM) and four different PtdIns(4,5)P_2 mole fractions (0.02, 0.05, 0.1, 0.2 mole fraction). Like the immunoprecipitated PLC-γ1, (His)$_6$-PLC-γ1 displays sigmoidal kinetics, (see *Figure 5* and *Table 1*). These data indicate that the model generated from kinetic data on the immunoprecipitated enzyme, and that there at are least two cooperative binding sites for lipid, are also valid for (His)$_6$-PLC-γ1. Using the computer software program UltraFit (BioSoft), kinetic parameters can be calculated (*Table 1*). As expected, the (His)$_6$-PLC-γ1 has a higher V_{max} than immunoprecipitated PLC-γ1 (5.56 μmol min^{-1} mg^{-1} versus 0.012 μmol min^{-1} mg^{-1} (23)) (*Table 1*). The $S_{0.5}$ value of the (His)$_6$-PLC-γ1 is lower than that of the immunoprecipitated form of PLC-γ1 (0.1 versus 0.28 mole fraction) (23). (His)$_6$-PLC-γ1 has a larger cooperativity index then the immunoprecipitated form of the enzyme (3.72 versus 2.5) (23).

Table 1. Comparison of the kinetic parameters of baculovirus-expressed PLC-γ1 versus immunoprecipitated PLC-γ1

Kinetic parameter	(His)$_6$-PLC-γ1	Immunoprecipitated PLC-γ1[a]
Vmax	5.56 μmol min^{-1} mg^{-1}	0.012 μmol min^{-1} mg^{-1}
$S_{0.5}$	0.1 mole fraction	0.32 mole fraction
n	3.72	2.5
Ks	0.03 mM	1.7 mM

[a] (23)

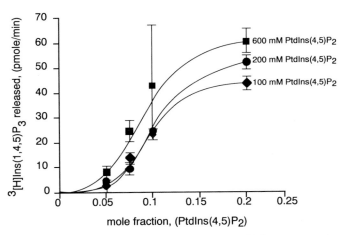

Figure 5. (His)$_6$-PLC-γ1 displays sigmoidal kinetics. (His)$_6$-PLC-γ1 was assayed at four mole fractions of PtdIns(4,5)P_2 (0.05, 0.075, 0.1, 0.2, 0.3) at three molar concentrations of PtdIns(4,5)P_2 (100, 200, 600 mM) using the detergent:PtdIns(4,5)P_2 mixed micelle assay system (*see Protocols 4 and 5*). Each point respresents the average of duplicate measurements. Each experiment was repeated four times with similar results.

4. *In vitro* phosphorylation of PLC-γ1

To investigate the effect of tyrosine phosphorylation on PLC-γ1 function it is desirable to tyrosine phosphorylate $(His)_6$-PLC-γ1. This can be performed *in vitro* by phosphorylating the purified $(His)_6$-PLC-γ1 using purified EGF receptor (see *Protocols 6 and 7*).

4.1 Isolation of A-431 membranes and purification of EGF receptors

A good source of cheap EGF receptor is A-431 membranes. A-431 cells contain high numbers of EGF receptors (5×10^6 receptors/cell), allowing the purification of high concentrations of functional receptor rapidly using simple techniques (32) (see *Protocol 6*). Further purification of EGF receptor can be carried out using affinity chromatography of WGA-sepharose purified EGF receptors on EGF-affigel column (33).

Protocol 6.   Purification of EGF receptor from A-431 membranes

Equipment and reagents

- A-431 cells
- Saponin (Sigma)
- Wheat-germ agglutinin conjugated to sepharose (WGA-sepharose) (Sigma)
- Solubilization buffer (20 mM HEPES, pH 7.4, 5% Triton X-100, 10% glycerol, 10 μg ml^{-1} leupeptin, 1 mM PMSF).
- 0.3 M N-acetyl-*D*-glucosamine (Sigma)

- Homogenization buffer (20 mM HEPES, pH 7.4, 5 mM KCl, 2 mM MgCl$_2$, 1 mM PMSF, protease inhibitors 1 μg ml^{-1} of leupeptin and aprotein)
- Equilibration buffer (20 mM HEPES, pH 7.4, 0.1 M NaCl, 0.2 % Triton X- 100, 10% glycerol, 10 μg ml^{-1} leupeptin)

Method

1. Preparation of A-431 membranes
   - (a) Solubilize 1×10^7 A431 cells in 2 ml of homogenization buffer containing 0.2% saponin.
   - (b) Lyse cells by 20 strokes with a teflon pestle in a mechanical homogenizer.
   - (c) Centrifuge the homogenate at 1000 rpm in a clinical centrifuge to pellet nuclei.
   - (d) Centrifuge the supernatant at 441 000 $\times$ *g* for 20 min to pellet membranes.
   - (e) Suspend membranes in 200 μl homogenization buffer.
   - (f) Store isolated membranes at $-80\,°C$.
2. Dilute membrane vesicles 1:5 with solubilization buffer^a.
3. Incubate membranes for 20 min at room temperature.
4. Remove particulate matter by centrifugation at 441 000 $\times$ *g* for 20 min.
5. Incubate solubilized membranes 1 ml equilibrated WGA-sepharose for 2 h.

Protocol 6. *Continued*

  (a) Equilibrate WGA-sepharose column.

  (b) Pack a column with 2 ml slurry of WGA-sepharose (WGA-sepharose) (Sigma).

  (c) Wash the column with 10 column volums of equilibration buffer.

6. Wash column with 20 column volums of equilibration buffer.

7. Elute EGF receptors from column with 1.0 ml equilibration buffer containing 0.3 M N-acetyl-D-glucosamine.

8. Aliquot and store EGF receptor at −80°C.

[a]Final Triton X-100 concentration should be 1% Triton X-100 per mg of protein.

4.2 *In vitro* phosphorylation of (His)$_6$-PLC-γ1

The WGA-sepharose purified EGF receptors can now be used to tyrosine phosphorylate (His)$_6$- PLC-γ1 (see *Protocol 7*). The addition of EGF to activate EGF receptor is not required, because WGA-sepharose purified EGF is activated. The addition of 1 mM DTT to the reaction mixture is essential to obtain good incorporation of phosphate into (His)$_6$-PLC-γ1. The amount of phosphate incorporated into (His)$_6$-PLC-γ1 can be monitored by the inclusion of [^{32}P]ATP to the reaction mixture. For some applications it is desirable to ensure the removal of EGF receptor from (His)$_6$-PLC-γ1. This can be achieved by purifying the tyrosine phosphorylated (His)$_6$-PLC-γ1 over Ni^{2+}-NTA agarose as described previously (*Protocol 2*)

Protocol 7. Phosphorylation of (His)$_6$-PLC-γ1

Equipment and reagents

- Phosphorylation reaction buffer (20 mM HEPES, pH 7.4, 10 mM MnCl$_2$, 0.2 mM Na$_3$VO$_4$, 10 mM ATP, 50 ng ml^{-1} BSA, 1 mM DTT)
- Stop buffer (50 mM Na phosphate, pH 7.4, 0.5 mM EGTA, 0.5 mM Na pyrophosphate)
- Phosphotyrosine antibody conjugated to agarose (Upstate Biochemical Incorporated)
- 0.1 μg WGA-sepharose purified EGF receptor (see *Protocol 6*)

- Antiphosphotyrosine column wash buffer (20 mM HEPES, pH 7.4, 100 mM NaCl, 0.5 mM β-glycerolphosphate, 0.2 mM Na$_3$VO$_4$, 0.5 mM EGTA and 0.5 mM Na pyrophosphate)
- Antiphosphotyrosine column elution buffer (20 mM HEPES, pH 7.4, 100 mM NaCl, 0.5 mM β-glycerolphosphate, 0.2 mM Na$_3$VO$_4$, 0.5 mM EGTA and 0.5 mM Na pyrophosphate)
- 2 μg (His)$_6$-PLC-γ1 (see *Protocols 2 and 3*)

Method

1. Incubate 2 μg of (His)$_6$-PLC-γ1 with phosphorylation reaction buffer, 0.1 μg WGA- sepharose purified EGF receptor in a reaction volume of 50 μl.

2. Incubate the reaction for 60 min at 4°C.

3. Stop the reaction by the addition of 0.5 ml of stop buffer containing EGTA.

4. Separate phosphorylated (His)$_6$-PLC-γ1 from non-phosphorylated by affinity purification on antiphosphotyrosine column.

(a) Incubate the reaction mixture overnight (at 4°C with rocking) with 50 μl antiphosphotyrosine column matrix in a spin column.

(b) Wash the antiphosphotyrosine column matrix three times with 0.5 ml of antiphosphotyrosine column wash buffer containing phosphatase inhibitors.

(c) Elute (His)$_6$-PLC-γ1 from the antiphosphotyrosine column matrix by incubation with 50 μl of 20 mM phenylphosphate buffer containing phosphatase inhibitors and centrifugation at 7 000 × *g*.

Tyrosine phosphorylation of (His)$_6$-PLC-γ1 results in an increase in activity from 2.25 μmol min^{-1} mg^{-1} to 5.56 μmol min^{-1} mg^{-1}, a 2.5-fold increase. The magnitude of activation is similar to that observed to PLC-γ1 isolated by immunoprecipitation (15, 22, 23).

5. Conclusion

The development of the baculovirus expression system for production of PLC-γ1 protein allows the generation of milligram quantities of protein that can easily be purified to homogeneity by Ni^{2+} affinity chromatography. The purified (His)$_6$-PLC-γ1 is apparently fully functional; the enzyme is able to hydrolyse PtdIns(4,5)P_2 with high specific activity and displays kinetics similar to those previously described for immunoisolated PLC-γ1. The (His)$_6$-PLC-γ1 can also be tyrosine phosphorylated by the EGF receptor and tyrosine phosphorylation results in the increase in (His)$_6$-PLC-γ1 activity. The use of baculovirus expression system and the detergent:phospholipid mixed micelle will allow the evaluation of how various mutations of PLC-γ1 structural elements affect PLC-γ1 function.

References

1. Margolis, B., Rhee, S.G., Felder, S., Mervic, M., Lyall, R., Levitz, A., Ullrich, A., Zilberstein, A., and Schlessinger, J. (1989). *Cell*, **57**, 1101–1107.
2. Meisenhelder, J., Suh, P.-G., Rhee, S.G., and Hunter, T. (1989). *Cell*, **57**, 1109–1122.
3. Kim, H.K., Kim, J.W., Zilberstein, A., Margolis, B., Kim, C.K., Schlessinger, J., and Rhee, S.G. (1991). *Cell*, **65**, 435–441.
4. Rhee, S.G. and Choi, K.D. (1992). *Adv. Second Messenger and Phosphoprotein Research*, **26**, 35–61.
5. Wahl, M. and Carpenter, G. (1988). *J. Biol. Chem.*, **263**, 7581–7590.
6. Rhee, S.G. and Choi, K.D. (1992). *J. Biol. Chem.*, **267**, 12393–12396.
7. Wahl, M.I., Nishibe, S., Suh, P.-G., Rhee, S.G., and Carpenter, G. (1989). *Proc. Natl. Acad. Sci. USA*, **86**, 1568–1572.

8. Hepler, J.R., Jeffs, R.A., Huckle, W.R., Outlaw, H.E., Rhee, S.G., Earp, H.S., and Harden, T. K. (1990). *Biochem. J.*, **270**, 337–344.
9. Kim, U.H., Fink, D., Jr., Kim, H.S., Park, D.J., Contreras, M.L., Guroff, G., and Rhee, S.G. (1991). *J. Biol. Chem.*, **266**, 1359–1362.
10. Todderud, G., Wahl, M.I., Rhee, S.G., and Carpenter, G. (1990). *Science*, **249**, 296–298.
11. Nishibe, S., Wahl, M.I., Rhee, S.G., and Carpenter, G. (1989). *J. Biol. Chem.*, **264**, 10335–10338.
12. Diakonova, M., Payrastre, B., van Velzen, A.G., Hage, Willem J., van bergen en Henegouwen, P.M.P., Boonstra, J., Cremers, F.F.M., and Humber, B.M. (1995). *J. Cell Science*, **198**, 2499–2509.
13. Atkinson, T.P., Lee, C.-W., Rhee, S.G., and Hohman, R.J. (1993). *J. Immunol.*, **151**, 1448–1455.
14. Kim, J.W., Sim, S.S., Kim, U.H., Nishibe, S., Wahl, M.I., Carpenter, G., and Rhee, S.G. (1990). *J. Biol. Chem.*, **265**, 3940–3943.
15. Nishibe, S., Wahl, M.I., Hernandez-Sotomayor, S.M.T., Tonks, N.K., Rhee, S.G., and Carpenter, G. (1990). *Science*, **250**, 1253–1256.
16. Wahl, M.I., Olashaw, N.E., Nishibe, S., Rhee, S.G., Pledger, W.J., and Carpenter G. (1989). *Mol. Cell. Biol.*, **9**, 2934–43.
17. Wahl, M.I., Nishibe, S., Kim, J.W., Kim, H., Rhee, S.G., and Carpenter, G. (1990). *J. Biol. Chem.*, **265**, 3944–3948.
18. Horstman, D.A., Ball, R., and Carpenter, G. (1995). *Protein Expression and Purification*, **6**, 278–283.
19. Dennis, E.A. (1974). *J. Supramolec. Struct.*, **2**, 682–694.
20. Hendrickson, H.S. and Dennis, E.A. (1984). *J. Biol. Chem.*, **259**, 5734–5739.
21. Roberts, M.F., Deems, R.A., and Dennis, E.A. (1977). *Proc. Natl. Acad. Sci. USA*, **74**, 1950–1954.
22. Wahl, M.I., Jones, G.A., Nishibe, S., Rhee, S.G., and Carpenter, G. (1992). *J. Biol. Chem.*, **267**, 10447–10456.
23. Jones, G.A. and Carpenter, G. (1993). *J. Biol. Chem.*, **268**, 20845–20850.
24. Suh, P.-G., Ryu, S.H., Choi, W.C., Lee, K.-Y., and Rhee, S.G. (1988). *J. Biol. Chem.*, **263**, 14497–14504.
25. Ryu, S.H., Cho, K.S., Lee, K.-Y., Suh, P.-G., and Rhee, S.G. (1987). *J. Biol. Chem.*, **262**, 12511–12518.
26. Goldschmidt-Clermont, P.J., Machesky, L.M., Baldassare, J.J., and Pollard, T.D. (1990). *Science*, **30**, 1575–1577.
27. Goldschmidt-Clermont, P.J., Kim, J.W., Machesky, L.M., Rhee, S., and Pollard, T.D. (1991). *Science*, **251**, 1231–1233.
28. Lin, L.P. and Carman, G.M. (1990). *J. Biol. Chem.*, **265**, 166–170.
29. Diez, E., Louis-Flamberg, P., Hall, R., and Mayer, R.J. (1992). *J. Biol. Chem.*, **267**, 18342–18348.
30. Bae-Lee, M. and Carman, G.M. (1990). *J. Biol. Chem.*, **265**, 7221–7226.
31. James, S.R., Paterson, A., Harden, T.K., and Downes, C.P. (1995). *J. Biol. Chem.*, **270**, 11872–11881.
32. Carpenter, G., King, L., Jr., and Cohen, S. (1979). *J. Biol. Chem.*, **254**, 4884–4891.
33. Cohen, S., Carpenter, G., and King, L., Jr. (1980). *J. Biol. Chem.*, **255**, 4834–4842.

5

Baculovirus-promoted expression, purification, and functional assay of G-protein regulated PLC-β isoenzymes

ANDREW PATERSON, THERESA M. FILTZ, and
T. KENDALL HARDEN

1. Introduction

Knowledge of the catalytic and regulatory activities of inositol phospholipid-specific phospholipases C (PLC) is critical to understanding at a molecular level the transmembrane signalling response to many hormones, neurotransmitters, and growth factors. These ubiquitous enzymes catalyse release of the second messengers inositol 1,4,5-trisphosphate ($Ins(1,4,5)P_3$) and sn-1,2-diacylglycerol (DAG). Three structural families of mammalian PLC isoenzymes (PLC-β, -γ, and -δ) have been identified, and the PLC-β class of isoenzymes accounts for the signal transduction response to the largest group of hormones and neurotransmitters.

Members of the PLC-β family of isoenzymes are subject to multiple mechanisms of regulation. The most clearly described is the activation by GTP bound α-subunits of the pertussis toxin-insensitive Gq class of G-proteins (heterotrimeric guanine nucleotide-binding regulatory proteins) and by G-protein βγ-subunits. Regulation of PLC-β isoenzymes by G-protein subunits has been demonstrated both by expression of heterologous proteins in cultured mammalian cells and by reconstitution of purified proteins *in vitro*. This chapter describes *in vitro* quantitation of PLC-β activity in both the absence and presence of G-protein subunits. Additionally, schemes are described for purification of PLC-β isoenzymes after their expression in cultured insect cells using recombinant baculovirus. These expression, purification, and assay protocols are straightforward and can be accomplished in most laboratories without specialized equipment.

2. Measurement of PLC-β-catalysed inositol lipid hydrolysis

2.1 Assay of PLC-β activity in the absence of G-protein subunits

The reaction catalysed by PLC-β can be measured routinely in the absence of G-protein subunits with PtdIns(4,5)P_2/cholate vesicles (1). This assay facilitates monitoring of PLC-β during purification but is applicable to other tasks, such as assessment of the dependence of the PLC-β-catalysed reaction on free calcium. This method is applicable to all PLC-β isoenzymes and is employed routinely within this laboratory for assay of PLC-β1, PLC-β2, PLC-βt from avian erythrocytes, and PLC-βX from *Xenopus laevis*. Phosphatidylinositol 4,5-bisphosphate (PtdIns(4,5)P_2) or phosphatidylinositol 4-monophosphate are equally appropriate as substrates and preparation of both lipids from bovine brain extract (Folch fraction I) and from [^{3}H]inositol-labelled turkey erythrocytes is described elsewhere (2). The assay is sensitive to detection of 1 ng of PLC-β1. The rate of reaction is linear over the range 0–25 pmol min^{-1} with respect to both time (0–20 min) and enzyme concentration. The time of incubation at 30°C should be limited such that less than 30% of substrate is hydrolysed. Dilution of PLC-β in 10 mmol l^{-1} Hepes pH 7.2 may be necessary to maintain quantitative rates of PtdIns(4,5)P_2 hydrolysis. The above method provides a reaction mixture with PtdIns(4,5)P_2 available at a bulk concentration of 50 μmol l^{-1}. This value can be increased to 200 μmol l^{-1} and the reaction time limited to 2–5 min as an alternative strategy for maintenance of linear rates of reaction in the presence of high concentrations of PLC-β. The reaction velocity of PLC-β is dependent on free calcium. The described conditions provide a final concentration of approximately 100 μmol l^{-1} free calcium in the reaction mixture. However, assay of PLC-β in the presence of other components that bind calcium requires an increase in the concentration of CaCl$_2$ in the 4X assay buffer. Since high concentrations of Ca^{2+} do not inhibit PLC-β the concentration of CaCl$_2$ in the 4X assay buffer can be raised to 12.0 mmol l^{-1} in assays where proteins that bind calcium are likely to be encountered.

Protocol 1. Assay of PLC-β in presence of cholate

Equipment and reagents

- Chloroform/methanol stocks of PtdIns(4,5)P_2 and Ptd[^{3}H]Ins(4,5)P_2
- 4X assay buffer (10 mmol l^{-1} Hepes pH 7.2, 480 mmol l^{-1} KCl, 40 mmol l^{-1} NaCl, 8 mmol l^{-1} EGTA, 23.2 mmol l^{-1} MgSO$_4$, 8.4 mmol l^{-1} CaCl$_2$)
- 10 mmol l^{-1} Hepes pH 7.2

- 2% cholate in 10 mmol l^{-1} Hepes pH 7.2: sodium cholate is purified as described elsewhere (3) and stored as a 10% (w/v) solution
- Probe sonicator (Virsonic 50, Virtis Company)

Method

1. For 50 assays, mix 250 nmol of PtdIns(4,5)P_2 with Ptd[^{3}H]Ins(4,5)P_2 (750 000 c.p.m.) in a polypropylene tube and evaporate the solvent under a stream of nitrogen.
2. Resuspend the dried lipid in 1250 μl 10 mmol l^{-1} Hepes pH 7.2 by sonication (2 × 15 sec) and store on ice until use.
3. Mix 25 μl of sonicated lipid with 25 μl of 4X assay buffer and 25 μl of 2% cholate. Initiate reaction by adding PLC-β in 25 μl 10 mmol l^{-1} Hepes pH 7.2, and warm to 30°C in a water bath.
4. Terminate reaction after 15 min by addition of 375 μl of chloroform:methanol:HCl (40:80:1).
5. Add 125 μl chloroform and 125 μl 100 mmol l^{-1} HCl, vortex, and separate phases by centrifugation (2000 × g, 5 min, room temperature).
6. Remove 400 μl of upper phase and quantitate release of [^{3}H]Ins(1,4,5)P_3 by scintillation counting.

2.2 Kinetic analyses of PLC-β

The dependence of reaction velocity on bulk PtdIns(4,5)P_2 concentration is often hyperbolic and apparent values for V_{max} and K_m can be derived. However, the limited solubility of phospholipid substrate in bulk solution and the near saturating local concentration of PtdIns(4,5)P_2 within the vesicle prevents calculation of true values for V_{max} and K_m. Kinetic models are available that are appropriate for assays using detergent/phospholipid mixed micelles. Originally applied to small molecular weight PLA$_2$ (4), these models have been applied to a wide range of phospholipid-metabolising enzymes including PLC-β, PLC-γ, and PLC-δ. A fuller account of the Triton X-100/PtdIns(4,5)P_2 mixed micelle assay system and its application to PLC-γ is given elsewhere in this volume (Chapter 4). Approaches for derivation of values for interfacial K_m, V_{max}, and the micellar dissociation constant, K_s, for PLC-β are reported elsewhere (5).

2.3 Functional reconstitution of PLC-β with Gα$_{11}$

AlF$_4^-$-dependent PLC activity can be measured upon reconstitution of PLC-β with Gα$_{11}$, Gα$_q$, or a mixture of the two. The method involves mixing of G$_q$ α-subunit with cholate/phospholipid vesicles followed by removal of detergent by dialysis (6). Since this reconstitution method produces Gq α-subunit-containing phospholipid vesicles that are essentially detergent-free, problems associated with the effects of detergent on activation by Gq α-subunits (7, 8) are circumvented.

This method is routinely employed within the laboratory for assessment of concentration dependence of PLC-β isoenzymes on Gα$_{11}$ for activation. This is achieved by varying the final concentration of $G\alpha_{11}$ in the reaction mixture over the range $0.01-3$ nmol l^{-1}.

Although the above method employs $G\alpha_{11}$ purified from turkey erythrocyte plasma membranes (9), it also is applicable to reconstitution of PLC-β1 with $G\alpha_q$ or $G\alpha_{16}$. Purification of $G\alpha_q/G\alpha_{11}$ from bovine liver (10) and brain (11) has been described elsewhere, as has purification of homogeneous preparations of $G\alpha_q$, $G\alpha_{11}$, and $G\alpha_{16}$ after expression of recombinant protein (7, 12). Rapid purification of recombinant Gq α-subunits by immobilized metal ion affinity chromatography recently has been reported (13, 14). Similarly, the method is equally applicable to reconstitution of PLC-β2, PLC-βt from avian erythrocytes, and PLC-βX from *X. laevis* with Gq α-subunits.

Protocol 2.   Reconstitution of PLC-β with Gq α-subunits

Equipment and reagents

- Chloroform/methanol stocks of PtdIns(4,5)P_2 and Ptd[^{3}H]Ins(4,5)P_2 (*Protocol 1*); chloroform/methanol stocks of phosphatidylethanolamine and phosphatidylserine (Avanti Polar Lipids)
- Dialysis buffer (20 mmol l^{-1} Hepes pH 7.4, 2 mmol l^{-1} MgCl$_2$, 100 mmol l^{-1} NaCl, 2 mmol l^{-1} DTT, 100 μmol l^{-1} PMSF, 100 μmol l^{-1} benzamidine)
- Dilution buffer (20 mmol l^{-1} Hepes pH 7.4, 0.8% (w/v) sodium cholate, 1 mmol l^{-1} MgCl$_2$, 100 mmol l^{-1} NaCl, 2 mmol l^{-1} DTT, 100 μmol l^{-1} PMSF, 100 μmol l^{-1} benzamidine)

- $G\alpha_{11}$ purified from turkey erythrocyte plasma membranes
- 4X assay buffer (150 mmol l^{-1} Hepes pH 7.0, 300 mmol l^{-1} NaCl, 16 mmol l^{-1} MgCl$_2$, 8 mmol l^{-1} EGTA, 6.4 mmol l^{-1} CaCl$_2$, 2 mg ml^{-1} fatty acid-free bovine serum albumin)
- 10X AlF$_4^-$ (200 μmol l^{-1} AlCl$_3$, 100 mmol l^{-1} NaF. Prepared by 5-fold dilution of 1 mmol l^{-1} AlCl$_3$, 500 mmol l^{-1} NaF in 20 mmol l^{-1} Hepes pH 7.0 immediately before use)
- 20 mmol l^{-1} Hepes pH 7.0
- Probe sonicator (*Protocol 1*)

Method

1. For 15 assays mix 30 nmol PtdIns(4,5)P_2, Ptd[^{3}H]Ins(4,5)P_2 (225 000 c.p.m.), 120 nmol phosphatidylethanolamine, and 30 nmol phosphatidylserine in a polypropylene tube and evaporate the solvent under a stream of nitrogen.

2. Resuspend the dried lipid in 150 μl dilution buffer by sonication (2 $\times$ 15 sec) and place on ice. Avoid foaming.

3. Dilute 150 ng $G\alpha_{11}$ to 25 μl with dilution buffer, mix with sonicated lipid, and adjust to 300 μl with dilution buffer.

4. Dialyse 14–20 h against 2 litres dialysis buffer[a].

5. Collect dialysed vesicles, dilute to 750 μl with dialysis buffer, and place on ice.

6. Mix 25 μl 4X assay buffer , 50 μl dialysed vesicles, 5 μl 20 mmol l^{-1} Hepes pH 7.0 , and 10 μl 10X AlF$_4^-$.[b]

7. Initiate reaction with 1–10 ng PLC-β1 diluted in 10 μl 20 mmol l^{-1} Hepes pH 7.0 , and warm to 30 °C in a water bath.

8. Terminate reaction after 5–15 min by addition of 375 μl chloroform: methanol:HCl (40:80:1).

9. Add 125 μl chloroform and 125 μl 100 mmol l^{-1} HCl, vortex, and separate phases by centrifugation (2000 × g, 5 min, room temperature).

10. Remove 400 μl of upper phase and quantitate release of [^{3}H]Ins(1,4,5)P_3 by scintillation counting.

[a]Dialysis of the cholate/phospholipid/Gα_{11} mixture is performed in a 15-well dialysis chamber (model D1451, MRA International) with a dialysis membrane rated at 14 000 MWCO. Fresh dialysis buffer is introduced to the reservoir of the chamber at approximately 100 ml h^{-1}. This apparatus has the advantage of allowing simultaneous dialysis of up to 15 independent G-protein/vesicle preparations. Any multi-well dialysis chamber intended for use with small volume samples is equally applicable.

[b]Hydrolysis of PtdIns(4,5)P_2 may proceed in the absence of AlF$_4^-$, especially with high concentrations of either Gα_{11} or PLC-β1. The velocity of this reaction can be determined by substituting 10 μl 20 mmol l^{-1} Hepes pH 7.0 for the 10X AlF$_4^-$ in the final reaction mixture. The Gα_{11}-stimulated rate of reaction then is determined by comparison of the rates measured in the presence and absence of AlF$_4^-$.

2.4 Other strategies for reconstitution of Gq α-subunits with PLC-β

Avian Gα_{11} was purified to homogeneity from turkey erythrocyte plasma membranes on the basis of its capacity to activate PLC-βt in the presence of AlF$_4^-$ (9). Purification was monitored by a rapid reconstitution assay. The presence of Gα_{11} in column fractions was detected by combining small portions of the cholate-containing column eluate with substrate phospholipid/cholate vesicles and the detergent removed rapidly by gel filtration over 5 ml columns of Sephadex G-50F. Purified PLC-βt then was introduced and the AlF$_4^-$-dependent rate of reaction determined at 30°C. The rapidity with which this type of assay can be accomplished has obvious advantages in lengthy purification protocols. Recovery of Gq and phospholipid is less than quantitative using this gel filtration procedure but remains adequate for monitoring Gq α-subunit activity during purification.

Another method for rapid and extremely sensitive assay by reconstitution involves dilution of Gq into a suspension of phospholipid vesicles (15). Small portions (maximally 2 μl) of column fractions containing Gq and sodium cholate are combined with detergent-free substrate phospholipid vesicles prior to addition of PLC-βt and AlF$_4^-$. To maintain the quantitative nature of this method it may be necessary to dilute small portions (1 μl) of column fractions in buffer containing 0.8% cholate (dilution buffer, *Protocol 2*) prior to assay.

The reconstitution methods described above do not reproducibly support GTPγS-dependent activation of PLC-β. Low GTPγS sensitivity probably results from the extremely low rate of GDP/GTP exchange of Gq in the absence of activated receptors (16). Nonetheless, the above protocols may be altered to achieve increased sensitivity to GTPγS. Pre-incubation of Gq-containing vesicles in the presence of GTPγS (1 mmol l^{-1}) prior to introduction of PLC-β may be successful (7, 12) and may obtain saturable PtdIns(4,5)P_2 hydrolysis with respect to Gq α-subunit concentration. However, the method requires detergent which may alter basal enzyme kinetics.

2.5 Reconstitution of PLC-β with G-protein $\beta\gamma$-subunits

The PLC-β isoenzymes also are activated by G-protein $\beta\gamma$-subunits. This regulated activity can be observed *in vitro* with reconstitution technology similar to that employed for Gq α-subunits (6). Differential activation of the PLC-β isoenzymes is observed with this method (17, 18, 19). Differential activation of PLC-β1 to PLC-β2 does not result from disparate dependencies on concentration of $\beta\gamma$-subunit, but rather, it is due to the greater degree to which the reaction velocity catalysed by PLC-β2 is elevated (18).

The method in *Protocol 3* relies on removal of cholate by dialysis. We have no experience with methods allowing more rapid reconstitution. However, the detergent-dependent method applied to Gq α-subunits by others may be suitable for this purpose (7). Activation of PLC-β1 and PLC-β2 by $\beta\gamma$-subunits has been demonstrated using this alternative method, although the concentration dependencies on $\beta\gamma$-subunits for activation are greater than that measured with the above detergent-free method.

Protocol 3. Reconstitution of PLC-β with $\beta\gamma$-subunits

Reagents

- Chloroform/methanol stocks of PtdIns(4,5)P_2 and Ptd[^{3}H]Ins(4,5)P_2 (*Protocol 1*); chloroform/methanol stocks of phosphatidylethanolamine and phosphatidylserine (Avanti Polar Lipids)
- Dialysis buffer (20 mmol l^{-1} Hepes pH 7.4, 1 mmol l^{-1} MgCl$_2$, 100 mmol l^{-1} NaCl, 2 mmol l^{-1} DTT, 100 μmol l^{-1} PMSF, 100 μmol l^{-1} benzamidine)
- Purified bovine brain $\beta\gamma$-subunits

- Dilution buffer (20 mmol l^{-1} Hepes pH 7.4, 0.8% (w/v) sodium cholate, 1 mmol l^{-1} MgCl$_2$, 100 mmol l^{-1} NaCl, 2 mmol l^{-1} DTT, 100 μmol l^{-1} PMSF, 100 μmol l^{-1} benzamidine)
- 4X assay buffer (150 mmol l^{-1} Hepes pH 7.0, 300 mmol l^{-1} NaCl, 16 mmol l^{-1} MgCl$_2$, 8 mmol l^{-1} EGTA, 6.4 mmol l^{-1} CaCl$_2$, 2 mg ml^{-1} fatty acid-free bovine serum albumin)
- 20 mmol l^{-1} Hepes pH 7.0

Method

1. For 15 assays prepare sonicated phospholipid/cholate vesicles as described in steps 1 and 2 of the Gα_q/Gα_{11} reconstitution method (*Protocol 2*).

2. Dilute 1.5 μg $\beta\gamma$-subunits to 25 μl with dilution buffer, mix with 50 μl sonicated lipid, and adjust to 300 μl with dilution buffer.

3. Dialyse for 14–20 h against 2 litres dialysis buffer (see *Protocol 2*).

4. Collect dialysed vesicles, dilute to 750 μl with dialysis buffer, and place on ice until use.

---

5. Mix 25 μl 4X assay buffer, 50 μl dialysed vesicles, and 15 μl 20 mmol l^{-1} Hepes pH 7.0.

6. Initiate the reaction with 1–10 ng PLC-β2 in 10 μl 20 mmol l^{-1} Hepes pH 7.0, and warm to 30 °C in a water bath.

7. Terminate the reaction after 5–15 min by the addition of 375 μl chloroform:methanol:HCl (40:80:1) and determine the release of [^{3}H]Ins(1,4,5)P_3 as described in steps 9 and 10 for reconstitution of PLC-β1 with Gα_{11} (*Protocol 2*).

3. Expression and purification of recombinant PLC-β isoenzymes

Protocols for purification of PLC-β isoenzymes from native tissue sources are detailed elsewhere (e.g. refs. 8, 20, 21). They have required access to either PLC-β isoenzyme-selective immunoaffinity matrices (8) or large amounts of starting material (21). The yield of PLC-β typically has been low, i.e. micrograms, and preparations can be contaminated with splice variants (22), with proteolytic degradation products (23), and/or with other proteins. These problems are circumvented by purification of recombinant isoenzymes after their baculoviral-driven expression in Sf9 cells. Here we detail our current approaches for maintenance, infection, and harvest of Sf9 cells.

3.1 Growth and manipulation of Sf9 cells

3.1.1 Maintenance of Sf9 cells in serum-containing medium

Stock cultures of Sf9 cells (ATCC CRL-1711) are maintained as stirred suspensions. The cells will remain viable for several months in serum-containing medium with routine sub-culturing of the stock to clean, sterile, spinner flasks. For a more comprehensive review of growth, maintenance, and infection of Sf9 or other lepidopteran cells the reader is directed to one of several excellent manuals that are available currently (24).

Method

We recommend that a 250 ml spinner flask (Bellco #1965–00250) be modified by replacement of the impeller blade. A longer blade with only 4–5 mm clearance from the flask wall is recommended. This ensures maximum disruption of the liquid/air interface. The blade supplied for the 500 ml spinner flask (Bellco #A-523–199) is readily cut and adapted for this purpose. Alternatively, it may be possible to obtain spinner flasks with impeller blades of appropriate length from other manufacturers.

Sf9 cells revived from frozen stocks are grown for several passages as monolayer cultures in "TMN-FH" (24) comprised of Grace's Antherean Medium (Vaughn's modification, JRH Biosciences) supplemented with fetal bovine serum (10% v/v), Yeastolate (3.3 g l^{-1}), lactalbumin hydrosylate (3.3 g l^{-1}), penicillin (100 IU ml^{-1}), streptomycin (100 μg ml^{-1}), and amphotericin B (0.25 μg ml^{-1}). The L-glutamine is replenished on a frequent basis. All culture of Sf9 cells should be performed under aseptic conditions. Cells are adapted to suspension culture simply by scraping an overconfluent monolayer into 10–20 ml TMN-FH with a sterile cell scraper. The monolayers from two 175 cm^2 tissue culture flasks should be combined for this purpose. The suspension is pelleted (500 $\times$ g, 2 min), the supernatant aspirated, and the cell pellet resuspended to 60 ml with TMN-FH. The suspended cells are placed in a 100 ml spinner flask (Bellco #1965–00100) and maintained at 27 °C with an impeller speed of 70 r.p.m. The height of the impeller blade is adjusted to maximize disruption of the liquid/air interface. The initial cell density should be approximately 5 $\times$ 10^5 cells/ml, and the suspension is incubated until it reaches a density of approximately 2.5 $\times$ 10^6 cells/ml. The culture then is diluted to 5 $\times$ 10^5 cells/ml, the volume is maintained at 60 ml, and the excess cell suspension discarded. After two or three passages the suspension should be transferred in a total volume of 120 ml to a sterile 250 ml spinner flask. The suspension is stirred at 70 r.p.m. and the cell density maintained between 1 and 2.5 $\times$ 10^6 ml^{-1} by dilution every 36–48 h.

3.1.2 Adaptation to Serum-Free Media

Adaptation of suspension cultures of Sf9 cells to serum-free medium (Excell 401, JRH Biosciences, supplemented with 100 μg ml^{-1} streptomycin, 100 IU ml^{-1} penicillin, 0.25 μg ml^{-1} amphotericin B) is often advantageous. Purification of recombinant proteins from cells infected in the absence of serum may be simpler due to the absence of bovine serum proteins. However, the primary advantages include the decreased cost of serum-free medium and the increased cell viability that is experienced, particularly one or two days post-infection. Sf9 cells have been reported to exhibit lower levels of recombinant protein expression, and possibly viability, when maintained in serum-free medium for extended periods of time (24). For this reason it is essential to maintain stock cultures in serum-containing medium from which serum-free cultures can be derived as required.

Incremental adaptation of Sf9 cells to serum-free medium has been recommended (24). This is time-consuming; complete adaptation to 100% serum-free medium may take two weeks. However, we have found that Sf9 cells can be adapted over a period of a few days. Thus, 6 $\times$ 10^7 Sf9 cells growing in serum-containing medium are pelleted by centrifugation (500 $\times$ g, 2 min), resuspended into 120 ml serum-free medium, and seeded in a fresh, sterile 250 ml spinner flask. The cells are maintained at 27 °C and stirred at 70 r.p.m. until they reach a density of 2.5 $\times$ 10^6 ml^{-1}. The cells may begin to clump during this period. Lower the cell density to 5 $\times$ 10^5 ml^{-1} by dilution in serum-free medium and increase agitation to 110 r.p.m. The cell density should rise to 2.5 $\times$ 10^6 ml^{-1} in three days. At this stage the culture can be maintained by dilution to 1×10^6 ml^{-1}, used for infection with recombinant baculovirus, or expanded. Expansion of the culture is achieved by dilution to 400 ml at 5×10^5 ml^{-1} in 500 ml spinner flasks. When the cell density approaches 2.5×10^6 ml^{-1}, dilution is repeated until enough cells are available.

3.2 Purification of recombinant PLC-β1

3.2.1 Purification of recombinant PLC-β1 from small-scale cultures

For many projects, tens of micrograms of PLC-β1 may represent several years supply of enzyme. There is little need for specialized equipment, and procurement of such quantities of enzyme easily can be accomplished in most laboratories outfitted for the study of modern biochemistry. Purification of recombinant PLC-β1 (rPLC-β1) and PLC-β2 (rPLC-β2) from baculovirus-infected cells maintained as monolayer cultures has been described previously (25). However, the investigator already maintaining a stock culture of Sf9 cells for this purpose is equally able to purify rPLC-β1 from small suspension cultures of baculovirus-infected cells without need for additional dedicated equipment. Small suspension cultures have the added advantage of being less demanding on both incubators and investigators. Here we describe a method that will provide one to two hundred micrograms of purified rPLC-β1 from a 40 ml culture with an investment of as little as twenty five workhours. The method does not require access to expensive chromatography columns but rather depends on the disposable HiTrap columns distributed by Pharmacia LKB Biotechnology. We operate these columns with an FPLC chromatography system. However, this is not essential and any chromatography pump system that allows formation of linear or step gradients is equally applicable.

Protocol 4. Expression, and purification of recombinant PLC-β1—40 ml cultures

Equipment and reagents

- Excell 401 serum-free insect cell medium supplemented with penicillin (100 IU ml^{-1}), streptomycin (100 μg ml^{-1}), and amphotericin B (0.25 μg ml^{-1})
- Sf9 cells, adapted to supplemented Excell 401, and in exponential growth
- Recombinant baculovirus encoding PLC-β1 (1×10^8 p.f.u. ml^{-1})
- 100 ml microcarrier spinner flask (Bellco #1965-00100)
- Lysis buffer (20 mmol l^{-1} Hepes pH 7.2, 5 mmol l^{-1} MgCl$_2$, 2 mmol l^{-1} EGTA, 200 μmol l^{-1} benzamidine, 200 μmol l^{-1} PMSF, 2 μmol l^{-1} pepstatin A)

- 5 ml and 1 ml HiTrap Q, and 1 ml HiTrap Heparin prepacked columns (Pharmacia LKB Biotechnology)
- Chromatography buffer (25 mmol l^{-1} Hepes pH 7.2, 10 mmol l^{-1} NaCl, 2 mmol l^{-1} EGTA, 2 mmol l^{-1} EDTA, 2 mmol l^{-1} DTT, 200 μmol l^{-1} benzamidine, 200 μmol l^{-1} PMSF, 2 μmol l^{-1} pepstatin A, and 5 μg ml^{-1} leupeptin)
- Storage buffer (20% v/v glycerol, 25 mmol l^{-1} Hepes pH 7.2, 10 mmol l^{-1} NaCl, 2 mmol l^{-1} EGTA, 2 mmol l^{-1} EDTA, 2 mmol l^{-1} DTT, 200 μmol l^{-1} benzamidine, 200 μmol l^{-1} PMSF, 2 μmol l^{-1} pepstatin A, and 5 μg ml^{-1} leupeptin)

Protocol 4. *Continued*

Method

1. Collect 120×10^6 Sf9 cells by centrifugation ($500 \times g$, 2 min), aspirate the medium, resuspend the cell pellet to 37 ml in supplemented Excell 401, and transfer to a sterile spinner flask.

2. Add 3.6 ml viral stock to cells, obtaining a multiplicity of infection of 3 with a viral density of $\sim 1 \times 10^7$ p.f.u. ml^{-1}.

3. Incubate with constant stirring (110 r.p.m.) for 48 h at 27°C. The incubator should be humidified, and the side-arm caps of the flask loosened to allow gaseous exchange.

4. Collect the virally-infected cells by centrifugation ($500 \times g$, 5 min, 20°C), aspirate the medium, resuspend cells in ice-cold lysis buffer, and incubate the lysate on ice for 10 min.

5. Homogenize the lysate with 15 strokes in a loose-fitting Dounce homogenizer, and remove unbroken cells and nuclei by centrifugation ($500 \times g$, 5 min, 4°C)

6. Clarify low-speed supernatant by centrifugation ($105\,000 \times g$, 65 min, 4°C) and retain high-speed supernatant.

7. Mix high-speed supernatant with an equal volume of chromatography buffer and apply at 2 ml min^{-1} to a 5 ml HiTrap Q column. Elute rPLC-β1 at 5 ml min^{-1} with a 0–500 mmol l^{-1} gradient of NaCl in 150 ml of chromatography buffer. rPLC-β1 elutes between 200 and 250 mmol l^{-1} NaCl.

8. Dilute the rPLC-β1 pooled from the HiTrap Q column with an equal volume of chromatography buffer and apply at 1 ml min^{-1} to a 1 ml HiTrap Heparin column. Elute rPLC-β1 at 1 ml min^{-1} with a 0–1 mol l^{-1} gradient of NaCl in 30 ml of chromatography buffer. The rPLC-β1 elutes at approximately 400–450 mmol l^{-1} NaCl[a].

9. Dilute the rPLC-β1 pooled from the HiTrap Heparin column with 3 volumes of storage buffer and apply at 1 ml min^{-1} to a 1 ml HiTrap Q column. rPLC-β1 is eluted in a minimal volume at 1 ml min^{-1} by a 0–400 mmol l^{-1} step gradient of NaCl in storage buffer. The concentrated rPLC-β1 elutes in a volume of approximately 2 ml.

10. The purified enzyme is stable in excess of 1 year when stored at -80°C.

[a]The rPLC-β1 should be essentially homogeneous after HiTrap Heparin chromatography and the second HiTrap Q column serves solely to concentrate the enzyme. Incomplete resolution of rPLC-β1 from contaminant proteins on HiTrap Heparin may be experienced. This has been observed in the form of lower molecular weight species eluting in advance of rPLC-β1 in the NaCl gradient. The problem may be overcome simply: the rPLC-β1 eluting from HiTrap Heparin is pooled, diluted in at least an equal volume of chromatography buffer, and re-applied to the 1 ml HiTrap Heparin column reequilibrated in chromatography buffer. The rPLC-β1 is then eluted at 1 ml min^{-1} by a 0–400 mmol l^{-1} NaCl step gradient in chromatography buffer. The contaminants elute immediately as a sharp peak, whilst rPLC-β1 elutes isocratically as a broad peak of protein after approximately 4 ml of 400 mmol l^{-1} NaCl. The rPLC-β1 is readily concentrated by HiTrap Q chromatography as described in step 9 above.

3.2.2 Purification of recombinant PLC-β1 from large-scale cultures

The recombinant baculovirus/Sf9 expression system is equally adaptable to the production and purification of PLC-β isoenzymes in milligram amounts. Much information is already available concerning the need to maintain oxygen supply to large cultures of infected Sf9 cells (26, 27). Although these methods allow tens of litres of infected Sf9 cells to be maintained, they require substantial investment in specialized equipment. We have avoided this approach, and have enjoyed reasonable success with standard spinner flasks of the microcarrier type.

Protocol 5. Expression, and purification of recombinant PLC-β1—multiple litre cultures

Equipment and reagents

- Excell 401 serum-free insect cell medium supplemented with penicillin (100 IU ml^{-1}), streptomycin (100 μg ml^{-1}), and amphotericin B (0.25 μg ml^{-1})
- Sf9 cells, adapted to supplemented Excell 401, and in exponential growth
- 1 litre centrifuge bottles (Beckman Instruments #355676), sterilized by autoclaving (121°C, 15 p.s.i., 20 min)
- Recombinant baculovirus encoding PLC-β1 (1 × 10^8 p.f.u. ml^{-1})
- 500 ml or 1000 ml microcarrier spinner flasks (Bellco #1965-00500 or 1965-01000). Modify the impeller assembly of the 500 ml flask by replacing the magnet holder. We employ the magnet holder from a 250 ml Bellco flask (Bellco #A-523–201)
- Q-Sepharose FF, heparin Sepharose CL-6B, and 5 ml HiTrap Q column (Pharmacia LKB Biotechnology). Hydroxylapatite (BioGel HTP, BioRad Laboratories)

- Wash buffer (140 mmol l^{-1} NaCl, 40 mmol l^{-1} KCl, 1 mmol l^{-1} Na$_2$HPO$_4$, 10.5 mmol l^{-1} KH$_2$PO$_4$, pH 6.2)
- Lysis buffer (20 mmol l^{-1} Hepes pH 7.2, 5 mmol l^{-1} MgCl$_2$, 2 mmol l^{-1} EGTA, 200 μmol l^{-1} PMSF, 200 μmol l^{-1} benzamidine, 2 μmol l^{-1} pepstatin A)
- Chromatography buffer (25 mmol l^{-1} Hepes pH 7.2, 10 mmol l^{-1} NaCl, 2 mmol l^{-1} EGTA, 2 mmol l^{-1} EDTA, 200 μmol l^{-1} benzamidine, 200 μmol l^{-1} PMSF, and 2 μmol l^{-1} pepstatin A)
- Hydroxylapatite chromatography buffer (25 mmol l^{-1} Hepes pH 7.2, 10 mmol l^{-1} K$_2$HPO$_4$, 10 mmol l^{-1} KCl, 2 mmol l^{-1} DTT, 200 μmol l^{-1} PMSF, 200 μmol l^{-1} benzamidine, 2 μmol l^{-1} pepstatin A)
- Storage buffer (20 (v/v) glycerol, 25 mmol l^{-1} Hepes pH 7.2, 2 mmol l^{-1} EDTA, 2 mmol l^{-1} EGTA, 2 mmol l^{-1} DTT, 200 μmol l^{-1} PMSF, 200 μmol l^{-1} benzamidine, 2 μmol l^{-1} pepstatin A)

A. *Harvest of Sf9 cytosol containing rPLC-β1*

1. Under aseptic conditions transfer 3–4 × 10^9 Sf9 cells growing in log phase to autoclaved 1 litre centrifuge bottles. Collect the cells by centrifugation (JS-4.2 rotor, Beckman Instruments, 500 × *g*, 5 min, room temperature, brake set to zero).

2. Aspirate the supernatant and resuspend the cells to a density of 1.42 × 10^7 ml^{-1} in supplemented Excell 401. Add 0.9–1.2 × 10^{10} p.f.u. (60–120 ml) recombinant baculovirus encoding PLC-β1 and mix with gentle agitation. Incubate with occasional mixing for one h at room temperature. The conditions will result in a final cell suspension (1 × 10^7 cells/ml) infected at a multiplicity of virus of 3.

Protocol 5. *Continued*

3. Dilute the suspension to a density of 1.2×10^6 cells/ml and seed in 400 ml volumes in sterile 500 or 1000 ml spinner flasks. Incubate with constant stirring (110 r.p.m. for 500 ml flasks; 80 r.p.m. for 1000 ml flasks) for 48 h at 27°C. The height of the impeller blade is adjusted to maximize disruption of the air/liquid interface. The incubator should be humidified, and the side-arm caps of the flask loosened.

4. Collect the infected cells by centrifugation (see step 1 above). Resuspend the cell pellet to 1×10^7 ml^{-1} in wash buffer and collect by centrifugation.

5. Resuspend the washed cell pellet to 1×10^7 cells/ml in ice-cold lysis buffer, and incubate on ice for 15 min.

6. Homogenize the suspension with 15 strokes of a Dounce homogenizer (loose-fitting pestle) and remove the particulate material by centrifugation ($105\,000 \times g$, 65 min, 4°C).

7. Collect the supernatant, dilute in an equal volume of lysis volume containing 40% (v/v) glycerol and store at -80°C.

This is repeated twice until cytosol prepared from the equivalent of 9–12 $\times 10^9$ cells is available.

B. *Purification of rPLC-β1*

1. Thaw stored cytosol. Dilute with an equal volume of chromatography buffer, and apply at 6 ml min^{-1} to a 300 ml (50 × 150 mm) Q-Sepharose FF column. Wash the column with 1 litre of 100 mmol l^{-1} NaCl in chromatography buffer and elute rPLC-β1 with a 100–450 mM gradient of NaCl in 3.6 litres of chromatography buffer. rPLC-β1 elutes at approximately 250 mmol l^{-1} NaCl.

2. rPLC-β1 pooled from the Q-Sepharose FF column is diluted in an equal volume of chromatography buffer and applied at 6 ml min^{-1} to a 60 ml (26 × 120 mm) column of heparin–Sepharose CL-6B. Elute rPLC-β1 at 6 ml min^{-1} with a 100–1000 mmol l^{-1} gradient of NaCl in 1200 ml of chromatography buffer. rPLC-β1 will elute at approximately 470 mmol l^{-1} NaCl.

3. Dilute rPLC-β1 pooled from the heparin-Sepharose CL-6B in an equal volume of hydroxylapatite chromatography buffer and apply at 1 ml min^{-1} to a 40 ml (26 × 80 mm) column of BioGel HTP. Elute rPLC-β1 using a 100–600 mmol l^{-1} gradient of potassium phosphate (168 mmol l^{-1} KH$_2$PO$_4$ / 432 mmol l^{-1} K$_2$HPO$_4$) pH 7.2 in 600 ml of hydroxylapatite chromatography buffer. rPLC-β1 elutes between 450 and 600 mmol l^{-1} potassium phosphate.

4. Dilute rPLC-β1 pooled from the BioGel HTP column in 3 volumes of storage buffer and apply at 5 ml min^{-1} to a 5 ml HiTrap Q column . Wash the column with 25 ml storage buffer containing 100 mmol l^{-1} NaCl and elute the rPLC-β1 with a 100–1000 mmol l^{-1} step gradient of NaCl in the same buffer. Collect 1.0 ml fractions. The concentrated protein elutes in 3–5 ml. Store the purified protein at −80°C.

3.3 Purification of other recombinant PLC-β isoenzymes

The yield of recombinant PLC-β1 from 1.2×10^{10} Sf9 cells is approximately 30 mg. This method also is applicable to purification of recombinant PLC-β2, PLC-βt, and PLC-βX after their expression in Sf9 cells. Although the level of expression in Sf9 cells is equivalent to that observed with PLC-β1, the other PLC-β isoenzymes mainly partition with the particulate fraction upon hypotonic lysis. This is reflected in diminished yields of 6, 12, and 1.5 mg of purified PLC-β2, -βt, and -βX, respectively, that have been obtained with 1.2×10^{10} infected Sf9 cells. The yield of recombinant PLC-β2, -βt, -βX recovered in the soluble fraction is increased significantly if infected Sf9 cells are lysed by nitrogen cavitation leading to increased yields of purified phospholipase. For lysis by nitrogen cavitation, infected cells are resuspended to 1×10^{7} cells/ml in lysis buffer (*Protocol 5*), equilibrated to 500–600 p.s.i. for 30 min in a cell disruption bomb (model 4635, Parr Instrument Company), and discharged rapidly to atmospheric pressure.

The purity of PLC-β2, -βt, and -βX prepared by the protocol described above should be greater than 90%. Co-purification of a limited number of contaminants is observed. These can be removed either by anion-exchange chromatography (Mono Q) employing a 100–400 mmol l^{-1} linear gradient of NaCl in 20 column volumes or by incorporation of an ammonium sulphate precipitation step early in the protocol. For example, ammonium sulphate fractionation has been employed successfully with recombinant PLC-βt by slowly introducing concentrated ammonium sulphate solution to a solution of crude Sf9 cytosol containing PLC-βt. A concentration of ammonium sulphate reaching 40% saturation (equivalent to 243 g $(NH_4)_2SO_4$ added to 1 litre) is attained, the suspension incubated at 4°C for one hour, and the precipitate recovered by centrifugation. The precipitate is resuspended in chromatography buffer and PLC-βt recovered as soluble protein after centrifugation.

References

1. Morris, A.J., Waldo, G.L., Downes, C.P., and Harden, T.K. (1990). *J. Biol. Chem.*, **265**, 13 501.
2. Waldo, G.L., Morris, A.J., and Harden, T.K. (1994). In *Methods in enzymology* (ed. R. Iyengar). Vol. 238, p. 195. Academic Press, San Diego.

3. Waldo, G.L., Evans, T., Fraser, E.D., Northup, J.K., Martin, M.W., and Harden, T.K. (1987). *Biochem. J.*, **146**, 431.
4. Hendrickson, H.S. and Dennis, E.A. (1984). *J. Biol. Chem.*, **259**, 5734.
5. James, S.R., Paterson, A., Harden, T.K., and Downes, C.P. (1995). *J. Biol. Chem.*, **270**, 11 872.
6. Boyer, J.L., Waldo, G.L., and Harden, T.K. (1992). *J. Biol. Chem.*, **267**, 25 451.
7. Hepler, J.R., Kozasa, T., Smrcka, A.V., Simon, M.I., Rhee, S.G., Sternweis, P.C., and Gilman, A.G. (1993). *J. Biol. Chem.*, **268**, 14 367.
8. Smrcka, A.V. and Sternweis, P.C. (1993). *J. Biol. Chem.*, **268**, 9667.
9. Waldo, G.L., Boyer, J.L., Morris, A.J., and Harden, T.K. (1991). *J. Biol. Chem.*, **266**, 14 217.
10. Taylor, S.J., Smith, J.A., and Exton, J.H. (1990). *J. Biol. Chem.*, **265**, 17 150.
11. Pang, I.H. and Sternweis, P.C. (1990). *J. Biol. Chem.*, **265**, 18 707.
12. Kosaza, T., Hepler, J.R., Smrcka, A.V., Simon, M.I., Rhee, S.G., Sternweis, P.C., and Gilman, A.G. (1993). *Proc. Natl. Acad. Sci. USA.*, **90**, 9176.
13. Kosaza, T. and Gilman, A. (1995). *J. Biol. Chem.*, **270**, 1734.
14. Hepler, J.R., Biddlecombe, G.H., Kleuss, C., Camp, L.A., Hofmann, S.L., Ross, E.M., and Gilman, A.G. (1996). *J. Biol. Chem.*, **271**, 496.
15. Waldo, G.L., Boyer, J.L., and Harden, T.K. (1994). In *Methods in enzymology* (ed. R. Iyengar). Vol. 237, p. 182. Academic Press, San Diego.
16. Biddlecombe, G.H., Berstein, G., and Ross, E.M. (1996). *J. Biol. Chem.*, **271**, 7999.
17. Boyer, J.L., Graber, S.G., Waldo, G.L., Harden, T.K., and Garrison, J.C. (1994). *J. Biol. Chem.*, **269**, 2814.
18. Paterson, A., Boyer, J.L., Watts, V.J., Morris, A.J., Price, E.M. and Harden, T.K. (1995). *Cellular Signalling*, **7**, 709.
19. Waldo, G.L., Paterson, A., Boyer, J.L., Nicholas, R.A., and Harden, T.K. (1996). *Biochem. J.*, 316, 559.
20. Lee, K.-Y., Ryu, S.H., Pann-Ghill, S., Choi, W.C., and Rhee, S.G. (1987). *Proc. Natl. Acad. Sci. USA.*, **84**, 5540.
21. Jhon, D.-Y., Lee, H.-H., Park, D., Lee, C.-W., Lee, K.-H., Yoo, O.J., and Rhee, S.G. (1993). *J. Biol. Chem.*, **268**, 6654.
22. Bahk, Y.Y., Lee, Y.H., Lee, T.G., Seo, J., Ryu, S.H. and Suh, P.-G. (1994). *J. Biol. Chem.*, **269**, 8240.
23. Park, D., Jhon, D.-Y., Lee, C.-W., Ryu, S.-H., and Rhee, S.G. (1993). *J. Biol. Chem.*, **268**, 3710.
24. O'Reilly, D.R., Miller, L.K., and Luckow, V.A. (ed.) (1992). *Baculovirus expression vectors: a laboratory manual* (1st edn), p. 109. Freeman, New York.
25. Paterson, A. and Harden, T.K. (1996). In *Methods in neuroscience* (ed. P.C. Roche). Vol. 29. p. 246. Academic Press, Orlando.
26. O'Reilly, D.R., Miller, L.K. and Luckow, V.A. (ed.) (1992). *Baculovirus expression vectors: a laboratory manual* (1st edn), p. 241. Freeman, New York.
27. Rice, J.W., Rankl, N.B., Gurganus, T.M., Marr, C.M., Barna, J.B., Walters, M.M., and Burns, D.J. (1993). *BioTechniques*, **15**, 1052.

6

Purification and assay of PLC-δ

YOSHIMI HOMMA and YASUFUMI EMORI

1. Introduction

Phosphatidylinositol 4,5-bisphosphate (PtdIns(4,5)P_2) is a minor component of membrane phospholipids and functions as a precursor of intracellular second messengers. PtdIns(4,5)P_2 is hydrolysed by receptor-activated phospholipase C (PLC) to generate diacylglycerol and inositol 1,4,5-trisphosphate (Ins(1,4,5)P_3). Diacylglycerol is the physiological activator of protein kinase C, and Ins(1,4,5)P_3 activates Ca^{2+}-dependent systems through the release of Ca^{2+} from internal stores (1). Multiple members of the PLC superfamily have been identified in mammalian cells. These can be divided into three main subfamilies, PLC-β, PLC-γ, and PLC-δ, which are activated by different mechanisms (2, 3). PLC-β subfamily members have been identified as primary targets of heterotrimeric G proteins. Specifically, PLC-β1, PLC-β2, and PLC-β3 have been shown to be activated by α subunits of the Gq subfamily of G proteins and PLC-β2 and PLC-β3 are also activated by βγ subunits of some G protein species. On the other hand, it has been demonstrated that tyrosine kinases are involved in the activation of PLC-γ. Both PLC-γ1 and PLC-γ2 bind to growth factor receptors via their *src* homology 2 (SH2) domains in a ligand-dependent manner, and are then phosphorylated by these receptors, leading to activation. Recent reports indicate that non-receptor type tyrosine kinases also phosphorylate and activate PLC-γ.

In contrast to what is known about the PLC-β and PLC-γ subfamilies, the activation mechanism of PLC-δ remains unclear. Neither receptors nor transducers that couple to PLC-δ members have been identified. The PLC-δ subfamily contains so far four members: PLC-δ1, -δ2, -δ3, and -δ4. PLC-δ1 and PLC-δ2 can be purified from either rat or bovine brains (4–7). PLC-δ3 is known only at the cDNA level (8). PLC-δ4 cDNA and protein have been reported very recently (9). Structural analyses demonstrate that the molecular structure of PLC-δ contains three characteristic regions: the X and Y regions, which both show sequence similarity to the other PLC isoforms and are essential for enzymatic activity; a region which contains two EF hand-like motifs; and the pleckstrin homology (PH) region (*Figure 1*).

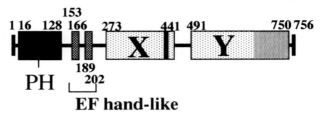

Figure 1. Schematic structure of PLC-δ. PLC-δ comprises the X and Y regions, two EF hand-like motifs, and the pleckstrin homology (PH) region. The amino acid residue numbers of each region are indicated. The activation mechanism of PLC-δ remains unclear.

We have recently cloned a novel regulator protein, p122, in the Rho-signalling pathway by screening a rat brain expression library with antiserum raised against purified PLC-δ1. This regulator protein shows a similarity to the GTPase-activating protein (GAP) homology region of Bcr and possesses GAP activity for RhoA, but not for Rac1; no guanine nucleotide-exchange activity for RhoA and Rac1 was detected. This novel p122RhoGAP binds to PLC-δ1 and activates its PtdIns(4,5)P_2 hydrolysing activity (10). These findings suggest that this novel RhoGAP is involved in the Rho-signalling pathway, probably downstream of Rho activation, and mediates the stimulation of PLC-δ. On the other hand, Rho and Rac have been shown to regulate a variety of cell functions, such as cell morphology, cell motility, cytokinesis, and platelet aggregation. These processes are triggered by a variety of extracellular stimuli including epidermal growth factor (EGF) and platelet-derived growth factor (PDGF). In addition, new functions for PtdIns(4,5)P_2 have been proposed since a number of PtdIns(4,5)P_2-binding proteins, such as gelsolin, cofilin, profilin, and α-actinin, have been found (11–17). These PtdIns(4,5)P_2-binding proteins are known to bind to actin and regulate actin assembly. Taken together, these findings suggest that PLC-δ members are involved in the regulation of various cell functions in which cytoskeletal organization is involved, such as cytokinesis and cell motility, through the hydrolysis of actin-regulated PtdIns(4,5)P_2.

In this chapter, we describe the procedures for purification and assay of PLC-δ1.

2. Purification of PLC-δ1 from rat brain

Several successful procedures for purification of PLC-δ1 have been reported. We describe here the method for purification from rat brains. Fresh brains are very good as starting materials, but brains frozen in liquid nitrogen and kept at −80°C for several months (from e.g. Bioproducts for Science, Indianapolis, IN, USA) are also good. All procedures are carried out in a cold room (~ 4°C). It is important that a purification series should finish within

4 days. About 10 μg of PLC-δ1 with 60–80% purity is obtained from 100 brains.

The method for extraction of PLC-δ1 is described in *Protocol 1*. The resultant supernatant is purified by performing a sequential series of chromatography runs (see *Figure 2*). Affi-gel blue column chromatography (*Protocol 2A*) is followed by heparin–sepharose column chromatography (*Protocol 2B*), gel filtration HPLC (*Protocol 2C*), HiTrap heparin HPLC (*Protocol 2D*) and then MonoS HPLC (*Protocol 2E*). The fractions obtained in the final step are assayed for PLC-δ1 using SDS-PAGE (*Figure 3*).

Protocol 1. Extraction of PLC-δ1 from rat brains

Equipment and reagents

- Polytron homogenizer (e.g. Brinkmann Instrument)
- PMSF: phenylmethylsulfonyl fluoride (e.g. Sigma P7626)
- DFP: diisopropyl fluorophosphate (e.g. Sigma D0879)
- Leupeptin (e.g. Sigma L2023)
- DTT: dithiothreitol (e.g. Sigma D5545)
- Homogenization buffer (20 mM Tris–HCl buffer, pH 7.5, 0.25 M sucrose, 2 mM EDTA, 0.5 mM EGTA, 1 mM PMSF, 1 mM DFP, 2 μg ml^{-1} leupeptin, 1 mM DTT)

Method

1. Wash 100 rat brains twice with 200 ml of ice-cold homogenization buffer.

2. Homogenize the brains three times (each for 1 min) with Polytron in 500 ml of ice-cold homogenization buffer.

3. Centrifuge the homogenate at 2000 × *g* for 15 min.

4. Centrifuge the resultant supernatant at 100 000 × *g* for 1 h.

5. Collect the supernatant as a starting material for sequential column chromatographies.

Protocol 2. Purification of PLC-δ1 by column chromatographies

Equipment and reagents

- HPLC apparatus (e.g. Pharmacia FPLC system)
- Apparatus and reagents for SDS–PAGE
- Apparatus and buffers for Western blotting
- Dialysis membrane (e.g. Spectrum CE/MWCO 8000)
- Buffers and coloring reagents for immunostaining (e.g. Promega W4100)
- Mes: 2-(*N*-morpholino)ethanesulfonic acid
- DTT (*Protocol 1*)
- Leupeptin (*Protocol 1*)
- Affi-Gel blue (Bio-Rad)
- Octyl-β-D-glucopyranoside (e.g. Sigma O8001)
- Polyethylene glycol 8000 (e.g. Sigma P2139)
- Anti-PLC-δ1 antibody (e.g. Santa Cruz, SASC-405)
- HiTrap heparin (Pharmacia)
- TSKgel G3000 SW (Tosoh)
- MonoS HR5/5 (Pharmacia)
- Heparin–Sepharose (Pharmacia)
- Column buffer A (10 mM Tris–HCl, pH 7.3, 1 mM EDTA, 10% (v/v) glycerol, 0.1 M NaCl, 2 μg ml^{-1} leupeptin)

Protocol 2. *Continued*

- Column buffer B (5 mM Mes-NaOH, pH 7.0, 1 mM EDTA, 10% (v/v) glycerol, 2 $\mu g\,ml^{-1}$ leupeptin)
- Column buffer C (5 mM Mes-NaOH, pH 6.5, 0.3% (w/v) octylglucopyronoside, 50 mM NaCl, 10% (v/v) glycerol, 1 $\mu g\,ml^{-1}$ leupeptin)
- Column buffer D (5 mM Mes-NaOH, pH 6.5, 10% (v/v) glycerol, 0.3% (w/v) octylglucopyronoside)

A. *Affi-Gel blue column chromatography*

1. Dilute brain extracts (*Protocol 1*) with 2 volumes of ice-cold 25 mM Tris–HCl (pH 7.5) buffer.

2. Apply the diluted sample to an Affi-Gel blue column (150 ml, 4 $\times$ 12 cm) equilibrated with column buffer A.

3. Wash the column with 1 litre of column buffer A containing 0.1 M NaCl.

4. Elute bound proteins with 300 ml of column buffer A containing 0.6 M NaCl.

B. *Heparin–Sepharose column chromatography*

1. Dialyse eluates against 1 litre of column buffer B three times.

2. Centrifuge dialysed sample at 100 000 $\times$ *g* for 1 h.

3. Apply the resultant supernatant to a heparin–Sepharose column (24 ml, 1.6 $\times$ 12 cm), previously equilibrated with column buffer B.

4. Wash the column with 200 ml of column buffer B.

5. Elute the column with a linear gradient from 0 to 0.5 M NaCl in 100 ml of column buffer B.

6. Collect 2 ml fractions.

7. Assay an aliquot of each fraction for PtdIns(4,5)P_2 hydrolysing activity (*Protocol 8*)

8. Detect PLC-δ1 protein by SDS–PAGE followed by immunoblotting with antibody agsainst PLC-δ1.

C. *Gel filtration HPLC*

1. Concentrate the peak fractions of PLC-δ1 from the heparin–sepharose column to 5 ml by reverse dialysis against 30% (v/v) polyethylene glycol 8000 in column buffer B.

2. Centrifuge the dialysed sample at 300 000 $\times$ *g* for 15 min.

3. Apply the resultant supernatant to a TSKgel G3000SW (2.6 $\times$ 40 mm) connected to an HPLC apparatus and previously equilibrated with column buffer C. Use a flow rate of 2 ml min^{-1}.

4. Collect 2 ml fractions.

5. Assay an aliquot of each fraction for PtdIns(4,5)P_2 hydrolysing activity (*Protocol 8*).

6. Detect PLC-δ1 protein by SDS–PAGE followed by immunoblotting with antibody against PLC-δ1.

D. HiTrap heparin HPLC

1. Apply peak fractions of PLC-δ1 from gel filtration column to a HiTrap Heparin (1 ml) connected to an HPLC system and previously equilibrated with column buffer D.

2. Wash the column with 5 ml of column buffer D.

3. Elute the column with a linear gradient from 0 to 0.5 M NaCl in 12.5 ml of column buffer D, using a flow rate of 0.5 ml min^{-1}.

4. Collect 0.5 ml-fractions.

5. Assay an aliquot of each fraction for PtdIns(4,5)P_2 hydrolysing activity (*Protocol 8*).

6. Detect PLC-δ1 protein by SDS–PAGE followed by immunoblotting with antibody agsainst PLC-δ1.

E. MonoS HPLC

1. Dialyse peak fractions of PLC-δ1 from the HiTrap heparin column twice against 200 ml of column buffer D.

2. Apply the sample to a Mono S HR5/5 column connected to an HPLC system and previously equilibrated with column buffer D.

3. Wash the column with 5 ml of column buffer D

4. Elute the column with a linear gradient from 0 to 0.5 M NaCl in 10 ml of the column buffer, using a flow rate of 0.5 ml min^{-1}.

5. Collect 0.5 ml-fractions.

6. Assay an aliquot of each fraction for PtdIns(4,5)P_2 hydrolysing activity (*Protocol 8*).

3. Production and purification of recombinant PLC-δ1 protein

In addition to purified material, we also use recombinant PLC-δ1 protein. Recombinant proteins produced by this protocol show Ca^{2+}-dependent activation of PtdIns(4,5)P_2 hydrolysis similar to that of purified PLC-δ1 (see results in *Figure 5*). This recombinant PLC-δ1 (rPLC-δ1) is activated by p122RhoGAP which binds to purified PLC-δ1. Therefore, rPLC-δ1 is a useful substitute for purified PLC-δ1 in various biochemical analyses.

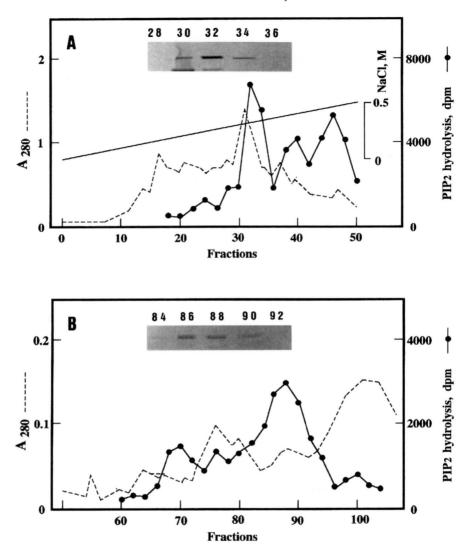

Figure 2. Purification of PLC-δ1 from rat brain by column chromatographies. (**A**) Heparin–Sepharose column chromatography. See *Protocol 2B* for details. (**B**) Gel filtration HPLC. Peak fractions of PLC-δ1 from the heparin–Sepharose column were pooled and concentrated, followed by gel filtration using TSKgel G3000SW HPLC. See *Protocol 2C* for details. (**C**) HiTrap heparin FPLC. Peak fractions of PLC-δ1 from the gel filtration column were pooled and applied to a HiTrap heparin column using an FPLC system. Bound proteins were eluted from the column with a linear NaCl gradient; 0.5 ml-fractions were

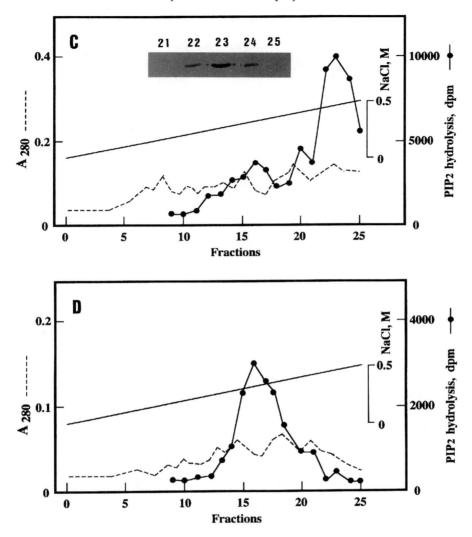

collected. See *Protocol 2D* for details. A 5 μl aliquot of each fraction was assayed for PtdIns(4,5)P_2 hydrolysing activity (*Protocol 8*). Inserts in (**A**), (**B**), and (**C**) show PLC-δ1 protein detected by SDS–PAGE followed by immunoblotting with antibody against PLC-δ1. (**D**) MonoS FPLC. Peak fractions of PLC-δ1 from the HiTrap heparin column were pooled and applied to a Mono S HR5/5 column using an FPLC system. See *Protocol 2E* for details.

Fractions

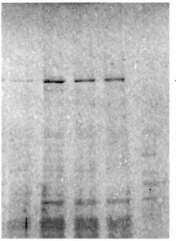

Figure 3. Results of purification of rat brain PLC-δ1. Samples of final preparations (20 μl of each fraction) were subjected to SDS–PAGE (6% acrylamide). The gel was stained with Coomassie Brilliant Blue. PLC-δ1 is indicated by an arrowhead.

3.1 Amplification of rat PLC-δ1 cDNA by PCR

When the cDNA clone covering the total coding sequence of PLC-δ1 is available, amplification of the cDNA is easily carried out using a pair of primers (see below and *Protocol 3* as an example) and *Pfu* DNA polymerase, a thermostable high fidelity DNA polymerase (*Protocol 3*). If not, it is also possible to obtain PLC-δ1 cDNA directly by the polymerase chain reaction (PCR), using tissue cDNA such as rat brain cDNA as a template. In this case, however, it is essential for efficient amplification to use *Taq* DNA polymerase (or its derivative) in place of *Pfu* DNA polymerase, although it may cause mutations. Hence the nucleotide sequences of obtained cDNA clones should be confirmed.

For PCR amplification 1 ng of rat PLC-δ1 cDNA is required, as well as 20 pmol each of the rat PLC-δ1-specific primers containing an appropriate restriction site (*Sma* I site in this case). The 5' primer has the sequence:

CCCCC<u>C</u>GG<u>G</u>**ATG**GACTCGGGTAGG

Bold letters denote the initiation codon of rat PLC-δ1, and underlined residues denote those different from the cDNA sequence and designed to generate *Sma* I site. The 3' primer has the sequence

CCACTCACTCAGGTCGGA

which corresponds to the antisense sequence of the 3'-noncoding region of PLC-δ1 cDNA and does not contain any restriction sites, because PLC-δ1 cDNA contains a single *Sma* I site just downstream of the termination codon.

A portion (5 μl) of the 50 μl reaction mixture should be checked by agarose gel electrophoresis for specificity of amplification, the amount and the length (approx. 2300 bp) of the product.

Protocol 3. PCR amplification of PLC-δ1 cDNA

Equipment and reagents

- Thermal cycler (e.g. Perkin Elmer)
- BSA: bovine serum albumin (e.g. Sigma A0281)
- *Pfu* DNA polymerase (2.5 U μl⁻¹, Stratagene)
- 10 × *Pfu* DNA polymerase buffer (100 mM KCl, 100 mM $(NH_4)_2SO_4$, 20 mM $MgSO_4$, 1% (w/v) Triton X-100, 1 mg ml⁻¹ BSA)
- 10 × dNTP solution (2 mM each of dATP, dCTP, dGTP, and TTP)

- The following 5'- and 3'-primers, specific for PLC-μ1 (20 pmol μl⁻¹):
  5': CCCCCGGGATGGACTCGGGTAGG
  3': CCACTCACTCAGGTCGGA
- PLC-δ1 cDNA covering the total coding sequence cloned into plasmid or phage vector (1 ng μl⁻¹)
- Apparatus and reagents for agarose gel electrophoresis (see ref. 18)

Method

1. Prepare in a 0.5-ml PCR tube the reaction mixture consisting of 5 μl 10 × *Pfu* DNA polymerase buffer, 5 μl 10 × dNTP solution, 1 μl each of the two primers (20 pmol each), 1 μl PLC-δ1 cDNA solution (1 ng), and 36 μl of water.

2. Add 1 μl (2.5 U) of *Pfu* DNA polymerase to each of reaction tubes, mix well, and then add one drop of mineral oil.

3. Centriguge the tubes for 5 sec and set into the thermal cycler.

4. Run the thermal cycler for 30 cycles; one cycle consists of denaturation at 94°C, annealing at 55°C, and polymerization at 72°C for 3 min.

5. Check the PCR products by subjecting an aliquot (5 μl) of the reaction mixture to agarose gel electrophoresis.

3.2 Construction of expression plasmid

If the expected DNA band is amplified by PCR, PLC-δ1 cDNA can be inserted into various expression vectors including bacterial, yeast and mammalian vectors among the numerous commercially available expression vectors. We have succeeded in the expression of PLC-δ1 using pGEX vectors (Pharmacia), glutathione S-transferase (GST) fusion protein vectors (*Figure 4*). Using pGEX-3X vector, PLC-δ1 cDNA amplified using the above primers should be inserted in-frame with the C-terminus of GST when the orientation is correct. Several clones carrying PLC-δ1 cDNA should be obtained and

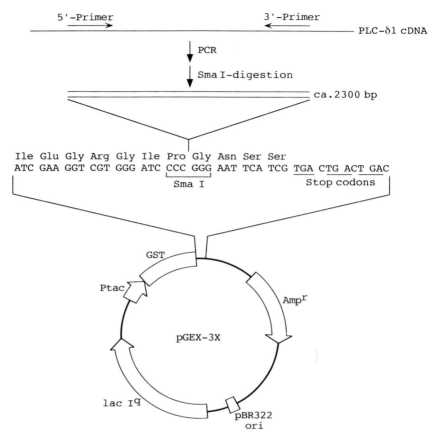

Figure 4. Construct of expression plasmid for rat PLC-δ1. Using the *Sma* I site of pGEX-3X vector, PLC-δ1 cDNA amplified by PCR and digested by *Sma* I is ligated to pGEX-3X to construct an expression plasmid for PLC-δ1. rPLC-δ1 is produced on a mg scale in *E. coli* AD202 (19) using this expression vector (*Protocol 5*).

their orientation and the nucleotide sequences bordering the ligation sites can be confirmed.

Protocol 4. Construction of an expression vector for a GST-PLC-δ1 fusion protein

Equipment and reagents
- Amplified DNA of PLC-δ1 (*Protocol 3*)
- BSA (*Protocol 3*)
- DTT (*Protocol 1*)
- *Sma* I
- pGEX-3X vector (Pharmacia)
- Bacto-trypton (Difco Laboratories)
- Bacto-yeast extract (Difco Laboratories)
- Ampicillin (e.g. Sigma A9518)

- Agar (e.g. Bacto-agar from Difco Laboratories)
- Bacterial alkaline phosphatase, 0.5 U μl^{-1} (e.g. Takara Shuzo)
- T4 DNA ligase, 350 U μl^{-1} (e.g. Takara Shuzo)
- pGEX 5' sequencing primer (Pharmacia)
- 1 × Sma I buffer (33 mM Tris-acetate, pH 7.9, 10 mM magnesium acetate, 0.5 mM DTT, 66 mM potassium acetate, 0.01% (w/v) BSA)
- 10 × T4 DNA ligase buffer (500 mM Tris–HCl, pH7.8, 100 mM MgCl$_2$, 100 mM DTT, 10 mM ATP, 500 μg ml^{-1} BSA)
- E. coli DH5α strain (or other E. coli strains showing high efficiency of transformation)
- Apparatus and reagents for agarose gel electrophoresis (see ref. 18)

- Phenol-chloroform (see ref. 18)
- Apparatus and reagents for plasmid preparation (see ref. 18)
- Apparatus and reagents for dideoxy nucleotide sequiencing (see ref. 18)

The following reagents should be sterilized by autoclaving:

- L-broth (10 g l^{-1} Bacto-trypton, 5 g l^{-1} Bacto-yeast extract, 10 g l^{-1} NaCl, pH 7.0–7.5)
- L-broth plate containing 1.5% (w/v) agar and 50 μg ml^{-1} (w/v) ampicillin (ampicillin should be added after cooling)
- Ca-Mg solution (80 mM CaCl$_2$, 50 mM MgCl$_2$)
- CaCl$_2$ solution (100 mM CaCl$_2$)
- Glycerol solution (50% (v/v) glycerol)

A. Processing of PCR product

1. Extract PCR products with phenol-chloroform (*Protocol 3*), and precipitate the DNA with ethanol. It is possible to purify the PCR products using a commercially available PCR product purification kit.

2. Dissolve DNA in 100 μl of 1 × Sma I buffer and then add Sma I (50 U) to the solution.

3. Incubate the solution at 30°C for 1–2 h.

4. Extract DNA with phenol-chloroform and precipitate it with ethanol.

5. Dissolve DNA in 45 μl water.

6. Check the amount and the length of the digested DNA by subjecting an aliquot (~1 μl) of the DNA solution to agarose gel electrophoresis.

B. Preparation of pGEX-3X vector for ligation

1. Dissolve pGEX-3X (1μg) in 100 μl of 1 × Sma I buffer and then add Sma I (50 U) to the solution.

2. Incubate the mixture at 30°C for 1–2 h.

3. Add 1 U bacterial alkaline phosphatase and 10 μl of 1 M Tris-HCl (pH 8.0) directly to the Sma I reaction mixture.

4. Incubate the mixture at 50°C for 30 min.

5. Extract DNA twice with phenol-chloroform, then precipitate it with ethanol.

6. Dissolve the DNA in 45 μl water.

7. Check the amount and the length of the digested DNA by subjecting an aliquot (~1 μl) of the DNA solution to agarose gel electrophoresis.

C. Ligation of two DNAs

1. Prepare four reaction tubes, each tube containing approx. 20 ng (estimated from the staining of the agarose gel) of the vector (*Step B* in

Protocol 4. *Continued*

this protocol), and 0, 20, 40, and 80 ng of PLC-δ1 cDNA (*Step A* in this protocol) in 20 μl of 1 × T4 DNA ligase buffer.

2. Add T4 DNA ligase (175 U) to each tube.

3. Incubate tubes at 15°C overnight.

D. *Preparation of competent E. coli*

1. Culture *E. coli* DH5α in L-broth overnight.

2. Inoculate 50 μl of the culture into 20 ml of fresh L-broth.

3. Incubate the fresh culture at 37°C with vigorous shaking until A_{550} reaches 0.4–0.45.

4. Chill the culture in ice-water.

5. Centrifuge the culture at 2000 × *g* for 10 min, and suspend the cell pellet in 2 ml of Ca–Mg solution.

6. After incubation for 5 min on ice, centrifuge the culture again at 2000 × *g* for 5 min, and suspend the pellet in 0.25 ml of $CaCl_2$ solution.

7. After incubation for 10 min on ice add 0.25 ml of glycerol solution, and use the resultant suspension as competent cells.

E. *Production of the expression plasmid*

1. To each ligation reaction tube prepared in *Step C*, add 80 μl of suspension of the competent cells prepared in *Step D*, and settle the tubes on ice for 30 min.

2. Incubate reaction tubes at 42°C for 55 sec and chill the tubes in ice-water.

3. Add 1 ml of L-broth to each tube, and incubate the tubes at 37°C for 30 min.

4. Centrifuge the tubes at 5000 × *g* for 1 min, and suspend cell pellet in 0.5 ml of L-broth.

5. Spread the cell suspension onto an L-broth plate containing 50 μg ml^{-1} ampicillin.

6. Incubate the plates at 37°C overnight to grow *E. coli* colonies.

(*Steps C* and *D* described above can be changed to the different transformation procedures in ref. 18)

7. Check the number of colonies (transformants), and pick up several colonies for plasmid preparation.

8. Select colonies carrying PLC-δ1 cDNA with the desired construct by the PCR, restriction map or colony hybridization (ref. 18) method.

9. Determine the nucleotide sequences bordering the GST-PLC-δ1 cDNA junction of the selected plasmids using pGEX 5' sequencing primer.

10. Confirm the construction of the expression plasmid.

3.3 Expression and purification of recombinant PLC-δ1

Using the expression vector constructed as described above, rPLC-δ1 is obtained on a mg scale in *E. coli* AD202 (19). This method is much more convenient than purification from rat brains as described in section 2 of this chapter. The synthesis of rPLC-δ1 can be induced by isopropyl-β-D-thiogalactopyranoside (IPTG) at the log phase of bacterial growth starting from fresh colonies. The temperature for cultures is important. It is better to culture *E. coli* at 30–32 °C rather than at 37 °C in order to obtain soluble and active rPLC-δ1.

rPLC-δ1 can be purified by glutathione Sepharose 4B essentially as a single band on SDS–PAGE. This fraction is ready to assay for PLC activity (*Protocol 8*). It is also possible to remove the GST portion by digestion with Factor X protease.

Protocol 5. Expression of recombinant GST-PLC-δ1 fusion protein (rPLC-δ1)

Equipment and reagents

- Expression plasmid of PLC-δ1 (*Protocol 4*)
- PBS: phosphate buffered-saline
- L-broth (*Protocol 4*)
- L-broth agar plate (*Protocol 4*)
- IPTG: isopropyl-β-D-thiogalactopyranoside (e.g. Sigma I6758)

- *E. coli* AD202 strain (ref. 19)
- Competent *E. coli* AD202 cells (prepared by *Protocol 4*)
- Reagents for *E coli* transformation (*Protocol 4*)

Method

1. Add 0.1 μg of the expression vector constructed in *Protocol 4* to the competent cells.

2. Incubate the reaction tubes at 42 °C for 55 sec and chill in ice-water.

3. Add 1 ml of L-broth to each tube and incubate the tubes at 37 °C for 30 min.

4. Centrifuge the tubes at 5000 × *g* for 1 min, and suspend the cell pellet in 0.5 ml of L-broth.

5. Spread the cell suspension onto an L-broth plate containing ampicillin.

6. Incubate the plates at 37 °C overnight to grow *E. coli* colonies.

7. Pick up several fresh colonies, and inoculate in 200 ml of L-broth containing 50 μg ml^{-1} ampicillin.

Protocol 5. *Continued*

8. Incubate cell suspensions at 37°C with shaking.
9. When A_{550} reaches ~0.4, add IPTG to give final concentration of 1 mM.
10. Incubate the cultures with vigorous shaking for an additional 2 h.
11. Centriguge the cultures at 2000 × *g* for 10 min at 4°C, then wash the cells with PBS.
12. Suspend the cell pellet in 2 ml ice-cold PBS.

Protocol 6. Purification of rPLC-δ1

Equipment and reagents

- Probe-type sonicator (e.g. Branson 250D Sonifire)
- Glutathione Sepharose 4B (Pharmacia)
- Factor Xa (e.g. Boehringer Mannheim)
- Sarcosyl solution (10% (w/v) sodium *N*-lauroyl sarcosinate)
- Triton solution (10% (w/v) Triton X-100)
- PBS (*Protocol 5*)
- PBS-Triton (PBS containing 0.2% (w/v) Triton X-100)
- Glutathione solution (50 mM Tris–HCl, pH 8.0, 10 mM glutathione, reduced form), freshly prepared
- Factor Xa cleavage buffer (50 mM Tris–HCl, pH 7.5, 150 mM NaCl, 1 mM $CaCl_2$)
- Factor Xa reaction solution (factor Xa cleavage buffer containing 50 μg ml^{-1} factor Xa)
- Apparatus and reagents for SDS–PAGE

Method

1. Add sarcosyl solution to the *E. coli* cell suspension (*Protocol 5*) to give a final concentration of 1% (w/v).
2. Sonicate the cells throughly with cooling with ice-water.
3. Add Triton solution to the sonicated *E. coli* to give 2% (w/v), and vortex the resultant solution vigorously.
4. Centrifuge the tubes at 15000 × *g* for 5 min.
5. Apply the supernatant to a glutathione Sepharose column (1 ml) previously equilibrated with PBS-Triton.
6. Wash the column twice with three colum volumes of PBS-Triton, once with two column volumes of PBS, and once with ten column volumes of factor Xa cleavage buffer.
7. Treat each millilitre of glutathione Sepharose column matrix with 1 ml of factor Xa reaction solution.
8. Replace the bottom cap on the column and allow the column to stand at room temparature overnight.
9. Remove the bottom cap and fractionate elutes of rPLC-δ1 protein into 0.3–0.5 portions.
10. Examine protein components of the eluates by SDS–PAGE, and collect the fractions containing rPLC-δ1.

4. Assay of PLC-δ1

The PtdIns(4,5)P_2 hydrolysing activity of PLC-δ1, like that of other PLC iso-forms, is measured by the formation of water-soluble Ins(1,4,5)P_3 from PtdIns(4,5)P_2. Although we describe here a conventional method for the assay of PtdIns(4,5)P_2 hydrolysis, a surface dilution method has been reported which seems to be more suitable for the kinetic analysis of PLC iso-forms than the conventional method (20). The PtdIns(4,5)P_2 hydrolysing activity PLC-δ1 is maximal at a submillimolar concentration of Ca^{2+} and therefore free Ca^{2+} concentrations should be adjusted to give the desired levels using a Ca^{2+}–EGTA buffer consisting of 3 mM EGTA and the appropriate amount of $CaCl_2$.

4.1 Preparation of PtdIns(4,5)P_2 substrate

The method for preparing labelled PtdIns(4,5)P_2 substrate is described in *Protocol 7*. With this method, it is possible to prepare PtdEth and PtdIns(4,5)P_2/[^{3}H]PtdIns(4,5)P_2 separately, and then combine them for the assay. It is also possible to eliminate PtdEth from the assay.

Protocol 7. Preparation of PtdIns(4,5)P_2 substrate

Equipment and reagents

- Stream of dry N_2
- Probe-type sonicator (*Protocol 6*)
- PtdIns(4,5)P_2 (Sigma P9763)
- Phosphatidyl [2–^{3}H]inositol [^{3}H]PtdIns(4,5)P_2(Du Pont-New England Nuclear)
- PtdEth: phosphatidylethanolamine (e.g. Sigma P8193)
- Dispersion buffer (5 mM Mes-NaOH, pH 6.5, 0.1 mM DTT)

Method

1. For 100 assays, mix 0.5 mg of PtdIns(4,5)P_2, 0.3 mg of PtdEth, and 74 kBq of [^{3}H]PtdIns(4,5)P_2 in a polypropylene tube and dry the mixture under a stream of dry N_2.

2. Add 1 ml of dispersion buffer to the lipids.

3. Sonicate the lipids mixture for 2 sec 5–10 times using probe-type sonicator. Keep the mixture on ice.

4. Centrifuge the lipid suspension at 10 000 × g for 10 min at 4°C.

5. Use 10 μl of supernatant as the lipid substrate for one assay.

4.2 Conventional assay for PtdIns(4,5)P_2 hydrolysis

Protocol 8 describes the conventional assay for PtdIns(4,5)P_2 hydrolysis. Under these conditions, [^{3}H]Ins(1,4,5)P_3 formation is linear with respect to time and enzyme concentration when less than 20% of substrate is consumed.

Protocol 8. Conventional assay for PtdIns(4,5)P_2 hydrolysis

Equipment and reagents

- Liquid scintillation counter
- Water bath-type incubater set at 30°C (or 37°C in case)
- 12 × 75 mm conical polypropylene tubes
- Liquid scintillation vials
- Substrate solutions
- 5 × assay buffer (100 mM Mes-NaOH, pH 6.5, 500 mM NaCl, 1 mg ml⁻¹ of BSA, 0.5 mM DTT)
- PLC source (e.g. an aliquot of chromatographic fraction)

- 5 × Ca^{2+}-EGTA buffer (15 mM EGTA and 30, 14, 12.5, 10, 6, 2.6, 0.95, and 0.03 mM CaCl$_2$ to yield 10^{-4} M, 10^{-5} M, 10$^{-5.5}$ M, 10^{-6} M, 10$^{-6.5}$ M, 10^{-7} M, 10$^{-7.5}$ M, and 10^{-8} M free Ca^{2+} concentrations respectively. For 0 M Ca^{2+}, 15 mM EGTA is used instead of Ca^{2+}-EGTA buffer.)
- Chloroform-methanol (1:1 by volume)
- 1 M HCl solution containing 1 mM EGTA
- Liquid scintillation cocktail for aqueous sample (e.g. PCS from Amersham)

Method

1. Prepare the assay mixtures in 12 × 75 mm conical polypropylene tubes on ice. The mixture (50 μl) contains 10 ml of lipid substrates, 10 μl of 5 × assay buffer, 10 μl of Ca^{2+}-EGTA buffer, 0–20 μl of enzyme source, and 20–0 μl of water.

2. Vortex the reaction mixture briefly and then incubate for 10 min at 30°C (or 37°C).

3. Add 2 ml of chloform/methanol (1:1 by volume) to a tube and vortex the mixture well.

4. Add 0.5 ml of 1 M HCl containing 1 mM EGTA to a tube and vortex the mixture well.

5. Centrifuge samples at 1500 × g for 3 min at room temparature.

6. Add a 700 μl portion of aqueous phase to 4 ml of scintillation fluid.

7. Subject scintillation samples to liquid scintillation counting.

8. Calculate the amount of Ins(1,4,5)P_3 from the radioactivity of [³H]Ins(1,4,5)P_3.

4.3 Reconstitution of PLC-δ1 with p122

Using this assay method it is also possible to compare the Ca^{2+}-dependent nature of the different enzymes and detect factors which affect PtdIns(4,5)P_2 hydrolysing activity of PLC-δ1. We introduce here two examples: (i) comparison of the Ca^{2+}-dependent PtdIns(4,5)P_2 hydrolysis of PLC-δ1 purified from rat brains and rPLC-δ1 (*Figure 5*), and (ii) activation of rPLC-δ1 by recombinant p122 (rp122) *in vitro* (*Figure 6*).

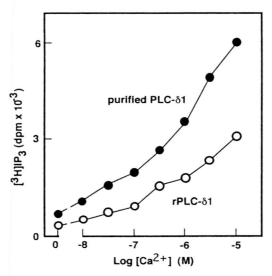

Figure 5. PtdIns(4,5)P_2 hydrolysing activity of purified PLC-δ1 and rPLC-δ1. The PtdIns(4,5)P_2 hydrolysing activity of either 5 pmol purified PLC-δ1 (closed circle) or 5 pmol rPLC-δ1 (open circle) was measured at the indicated concentrations of free Ca^{2+}. Assays were performed for 10 min at 30°C and released [^{3}H]Ins(1,4,5)P_3 was determined by scintillation counting (*Protocol 8*).

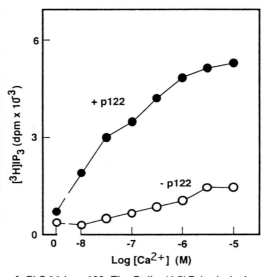

Figure 6. Activation of rPLC-δ1 by p122. The PtdIns(4,5)P_2 hydrolysing activity of 1 pmol rPLC-δ1 was determined at the indicated concentrations of free Ca^{2+} in the presence of about 1 pmol rp122 (closed circles) or 1 pmol GST (open circles). Assays were performed for 10 min at 30°C and released [^{3}H]IP$_3$ was determined by scintillation counting (*Protocol 8*).

The experimental procedures for assaying PtdIns(4,5)P_2 hydrolysing activity are the same as described above in section 4.2 and *Protocol 8*, except for the composition of reaction mixture. The mixture contains 10 μl lipid substrate, 10 μl assay buffer, 10 μl Ca^{2+}-EGTA buffers, 10 μl either purified PLC-δ1 or rPLC-δ1 protein solution, and 10 μl water or rp122 protein solution (10). The 5 × Ca^{2+}-EGTA buffer used consists of 15 mM EGTA and either 14, 12.5, 10, 6, 2.6, 0.95, or 0.03 mM CaCl$_2$ to give 10^{-5} M, 10$^{-5.5}$ M, 10^{-6} M, 10$^{-6.5}$ M, 10^{-7} M, 10$^{-7.5}$ M, or 10^{-8} M free Ca^{2+} concentrations respectively. For 0 M Ca^{2+}, 15 mM EGTA is used instead of Ca^{2+}-EGTA buffer.

References

1. Berridge, M.J. (1993). *Nature*, **361**, 315.
2. Rhee, S.G. and Choi, K.D. (1992). *Adv. Second Messenger Phosphoprotein Res.*, **26**, 35.
3. Sternweis, P.C. (1994). *Curr. Opin. Cell Biol.*, **6**, 198.
4. Ryu, S.H., Suh, P.-G., Cho, K.S., Lee, K.-Y., and Rhee, S.G. (1987). *Proc. Natl. Acad. Sci. U.S.A.*, **84**, 6649.
5. Suh, P.-G., Ryu, S.H., Moon, K.H., Suh, H.W., and Rhee, S,G. (1988). *Cell*, **54**, 161.
6. Meldrum, E., Katan, M., and Parker, P.J. (1989). *Eur. J. Biochem.*, **182**, 673.
7. Meldrum, E., Kriz, R.W., Totty, N., and Parker, P.J. (1991). *Eur. J. Biochem.*, **196**, 159.
8. Rhee, S.G., Lee, C.-W., and Choi, K.D. (1993). *Adv. Second Messenger Phosphoprotein Res.*, **28**, 57.
9. Lee, S.B. and Rhee, S.G. (1996). *J. Biol. Chem.*, **271**, 25.
10. Homma, Y. and Emori, Y. (1995). *EMBO J.*, **14**, 286.
11. Lassing, I. and Lindberg, U. (1985). *Nature*, **314**, 472.
12. Goldschumidt-Clermont, P.J., Machesky, L.M., Baldassare, J.J., and Pollard, T.D. (1990). *Science,* **247**, 1575.
13. Yonezawa, N., Nishida, E., Iida, K., Yahara, I., and Sakai, H. (1990). *J. Biol. Chem.*, **265**, 8382.
14. Yu, Y.-X., Johnston, P.A., Sudhof, T.C., and Yin, H.L. (1990). *Science*, **250**, 1413.
15. Banno,Y., Nakashima, T., Kumada, T., Ebisawa, K., Nonomura, Y., and Nozawa, Y. (1992). *J. Biol. Chem.*, **267**, 6488.
16. Fukami, K., Furuhashi, K., Inagaki, M., Endo, T., Hatano, S., and Takenawa, T. (1992). *Nature*, **359**, 150.
17. Janmey, P.A., Lamb, J., Allen, P.J., and Matsudaira, P.T. (1992) *J. Biol. Chem.*, **267**, 11 818.
18. Sambrook, I., Fritsch, E.F., and Maniatis, T. (ed.) (1989). *Molecular cloning: a loboratory manual* (2nd edn), p. 1. Cold Spring Harbor Laboratory Press, NY.
19. Nakano, H., Yamazaki, T., Ikeda, M., Masai, H., Miyatake, S., and Saito, T. (1993). *Nucleic Acid Res.*, **22**, 543.
20. Carman, G.M., Deems, R.A., and Dennis, E.A. (1995). *J. Biol. Chem.*, **270**, 18 711.

7

The purification and assay of inositide binding proteins

ANNE B. THEIBERT, GLENN D. PRESTWICH,
TREVOR R. JACKSON, and
LATANYA P. HAMMONDS-ODIE

1. Background

A valuable strategy towards understanding the physiological roles of inositol polyphosphates (Ins$P$ns) and inositol lipids (PtdIns$P$ns) will be the identification, characterization, isolation, and molecular biological and genetic manipulation of their intracellular receptors. Several experimental approaches have been used to isolate Ins$P$n and PtdIns$P$n binding proteins, including conventional chromatography and reversible ligand binding, Ins$P$n affinity column chromatography, and covalent modification of binding proteins with photoactivatable ligands. These approaches have led to the identification of an unexpected variety of Ins$P$n and PtdIns$P$n binding proteins, whose functions are only now being uncovered.

Many *myo*-inositol polyphosphates have been identified and studied in eukaryotic cells: InsP_6 and various isomers of InsP_3, InsP_4, and InsP_5 (summarized in Chapter 2). Levels of these isomers have been shown to be regulated differentially in diverse cell types, either increasing with acute agonist stimulation, varying during cell cycle progression, or remaining relatively constant. The Ins$P$n diphosphates PP-InsP_5 ('InsP_7') and [PP]$_2$-InsP_4 ('InsP_8') have also been identified from several cell types (see Chapters 2, 11 and refs. 1–4). Hence there exists a vast complexity of potential regulatory Ins$P$n molecules, some of which may function in acute signalling scenarios and some of which may function to regulate cell cycle and homeostatic activities. The expectation is that many of these Ins$P$ns, especially those whose levels are modulated, exert physiological effects that are mediated through the binding to and regulation of specific 'receptor' targets within the cell.

1.1 Ins(1,4,5)P_3

The first Ins$P$n shown to be an intracellular second messenger was Ins(1,4,5)P_3 (reviewed in (5), and see Chapter 10). The best characterized intracellular target for Ins(1,4,5)P_3 is its receptor calcium channel present on endoplasmic reticulum (ER) membranes (summarized in ref. 6). In addition to the ER Ins(1,4,5)P_3 receptor and related plasma membrane localized Ins(1,4,5)P_3 calcium channels (7–9), other potential targets for Ins(1,4,5)P_3 have also been identified either biochemically or physiologically, including voltage-gated calcium channels (10), an InsP_3-regulated protein phosphatase activity (11), phospholipase C$_\delta$ (12), and a calcium ATPase (13, 14). The identification of their physiological roles in phosphoinositide signalling awaits further characterization, which may be significantly accelerated by the identification and characterization of Ins$P$n binding motifs on these proteins.

There are a few reports of specific physiological activities of other InsP_3 isomers, including the putative nonphysiological isomers Ins(2,4,5)P_3 and Ins(1,2,6)P_3 (the latter is the so-called 'trinositol'). Interestingly, Ins(2,4,5)P_3 is substantially more potent than Ins(1,4,5)P_3 in binding to the rat olfactory Ins(1,4,5)P_3 receptor (15), raising the possibility that other InsP_3 isomers might also exert specific effects. Ins(1,2,6)P_3 has been shown to possess anti-inflammatory, anti-diabetic, anti-platelet, metal chelating, and vascular effects, reviewed in refs. 16 and 17. It has also been demonstrated to be an antagonist of neuropeptide Y induced vascular contraction *in vitro* and *in vivo* (18) and interferes with sympathetic neurotransmission (17). Whether this may reflect a function of an endogenous Ins$P$n awaits further investigation.

1.2 Other Ins$P$ns: more second messengers?

Many potential roles have been ascribed to the endogenous InsP_4s, InsP_5, and InsP_6. Ins(1,3,4,5)P_4 is formed rapidly after agonist stimulation by the phosphorylation of Ins(1,4,5)P_3 via a 3-kinase (for a review see ref. 19). Ins(1,3,4,5)P_4 has been implicated in the regulation of extracellular calcium entry into cells stimulated with agonists that increase Ins(1,3,4,5)P_4 production (summarized in ref. 20). In addition to the proposed role in regulating calcium entry into the cytoplasm, Ins(1,3,4,5)P_4 has been demonstrated to stimulate nuclear calcium influx (21). Furthermore, data from numerous electrophysiological studies suggest that Ins(1,3,4,5)P_4 plays a role in modulating potassium, chloride, and calcium ion fluxes in diverse cell types, either directly or indirectly through the regulation of intracellular calcium (see ref. 20 for a review, and also refs. 22–25). Data showing that Ins(1,3,4,6)P_4 and Ins(3,4,5,6)P_4 also regulate ion fluxes (26–28) suggest intracellular functions for these InsP_4 isomers as well.

Intracellular and extracellular functions for Ins(1,3,4,5,6)P_5 and InsP_6 have also been proposed (reviewed in refs. 29 and 30), including regulation of several cation conductances, calcium uptake, neurotransmitter release, and modulation of desensitization in agonist stimulated cells (31). InsP_6 displays anti-tumor activity (32), diminishes iron uptake into cells (33), primes neutrophils (34), and acts as an antagonist for numerous ligands, including the FGF family (35), P- and L-selectins (36) and IGF-II (37). InsPns have also recently been proposed to regulate synaptic vesicle fusion and/or recycling. Microinjection of InsP_4, InsP_5, and InsP_6 into squid giant axons inhibits neurotransmission (38). Roles for InsPns as a reservoir for inositol and phosphate, or as antioxidants, by their ability to chelate metal ions such as iron (particularly Fe^{3+}) (39) and so inhibit free radical production (40), have been suggested.

InsP_5 and InsP_6 exist in abundant, metabolically active pools that are converted to high-energy diphosphates (1–4). Functions for the inositol diphosphates have not yet been shown, but it appears clear that these high energy diphosphates are likely to play important roles in cellular physiology. PP-InsP_5 and [PP]$_2$-InsP_4 are implicated in ATP resynthesis from ADP (Chapter 11), in protein phosphorylation, and in synaptic vesicle recycling via high affinity binding to AP-2 (1–4, 41). These accumulating data demonstrate that many InsPns have potent, specific, intracellular and extracellular effects, supporting the existence of specific receptor binding proteins.

1.3 Physiological roles of PtdInsPns

Inositol lipids may also act as second messengers and are important regulators of protein function, such as membrane vesicle trafficking and cytoskeletal rearrangements (reviewed in refs. 42–45). PtdIns(4,5)P_2 has been demonstrated to modulate the activities of numerous proteins, including the small GTPase Arf and phospholipase D (46–48), and has also been suggested to regulate cytoskeletal dynamics by affecting the regulation of actin polymerization (43). PtdIns(4,5)P_2 has been shown to interact directly with pleckstrin homology (PH) domains in several proteins such as phospholipase C$_\delta$, α-actinin, dynamin, Grb-7, and β-spectrin (reviewed in refs. 49–51). Additionally, several other actin binding proteins, including profilin, gelsolin, vinculin, cofilin, villin, and CapZ, which do not contain conventional PH domains, have been shown to be regulated by PtdInsPns (reviewed in ref. 43).

Receptor stimulated PI 3-kinase activity and its product PtdIns(3,4,5)P_3 have been implicated in the regulation of numerous cellular activities, including mitogenesis (52), membrane ruffling and actin cytoskeletal reorganization (53–55), glucose transporter translocation (56–58), membrane vesicle trafficking (44, 59), and are thought to be involved in the stimulation of p70 ribosomal S6-kinase (58, 60), regulation of Akt kinase (PKB) (61–63), and neurite outgrowth (64, 65). Potential biochemical targets that mediate some of these physiological effects have also been identified. Both PtdIns(4,5)P_2 and PtdIns(3,4,5)P_3 activate several protein kinase C isoforms (66–71). Furthermore, several other proteins bind to or are regulated by 3-phosphorylated

PtdIns*P*ns. These include the regulatory p85 subunit of PI 3-kinase (72), a kinase responsible for pleckstrin phosphorylation (73), Akt kinase (PKB) (62), and PRK, a protein kinase C-related kinase (74). By analogy to the Ins*P*ns, the existence of specific receptors that mediate the effects of PtdIns*P*ns, especially PtdIns(3,4,5)P_3, has been proposed.

1.4 Ins*P*n binding in membrane fractions

As they can for InsP_3, high affinity binding sites for Ins(1,3,4,5)P_4 and InsP_6 can be measured in crude and enriched membrane fractions and by auto-radiography from many different tissues and species. Ins(1,3,4,5)P_4 binding sites were first identified in crude membranes from rat cerebellum and HL60 cells (75, 76), and have subsequently been characterized in crude membrane preparations from bovine adrenal cortex (77) and parathyroid (78), porcine and rat cerebellum (79, 80), rat brain and liver (81), colonic smooth muscle (82), and human platelets (83). Three different types of Ins(1,3,4,5)P_4 binding sites have been identified, as distinguished by acidic, neutral, and alkaline pH optima for binding (summarized in ref. 81). In platelet plasma membranes, an Ins(1,3,4,5)P_4 binding site with a K_d = 25 nM and a neutral pH optimum was identified (83). Identification of a high affinity Ins(1,3,4,5)P_4 binding sites (K_d = 1.6 nM) present on rat liver nuclear outer membranes was recently documented (21). Autoradiographic localizations in rat brain sections indicate that, under acidic binding conditions, [^{3}H]Ins(1,3,4,5)P_4 binding is specifically enriched in the hippocampus (84). [^{3}H]InsP_6 binding to brain, anterior pituitary, adrenal chromaffin, and retina membranes is both specific and of high affinity, with dissociation constants in the high nanomolar range (85–87). In autoradiography with [^{3}H]InsP_6 in rat brain sections, an enrich-ment of binding sites in the hippocampus and the granule cell layer of the cerebellum (84) was observed.

The molecular site of action of Ins(1,2,6)P_3 is also being actively studied. Modestly selective binding sites for Ins(1,2,6)P_3 have been measured in heart, aorta, and brain tissue (88). The structural similarity of this InsP_3 isomer with Ins(1,3,4,5)P_4 has led to speculation that these ligands compete for the same site (89). High specificity binding of labelled Ins(1,2,6)P_3 has been observed in human umbilical cord vascular endothelial cells (90). Using an Ins(1,2,6)P_3 photoaffinity label (see below) a 55 kDa umbilical cord epithelial cell target was identified (A. Chaudhary and G.D. Prestwich, unpublished data).

The original expectation was that each Ins*P*n would have perhaps one high affinity membrane binding site that would display a high degree of selectivity over other compounds, similar to what was determined initially for the InsP_3 receptor. Since the original binding studies were performed, many Ins*P*n binding proteins have now been identified, having varying affinities, specifici-ties, localizations, and levels of expression, making the previous data from crude membrane binding studies difficult to interpret. However, a reversible membrane binding assay can still be a valuable tool for studies of highly enriched membrane fractions. Purification of the membranes enriches for only a restricted set of specific binding proteins, facilitating the identification of specific InsPn receptors.

2. Identification

2.1 InsPn binding proteins

Several potential low affinity InsPn binding proteins have now been identified. Proteins that either bind InsPns or whose activities are affected by InsPns include avian erythrocyte hemoglobin (91), skeletal muscle aldolase (92), liver adenosine monophosphate deaminase (93), a brain protein phosphatase (11), and retinal arrestin (94). The functional significance of these interactions awaits further investigation and will likely depend on the characterization and genetic manipulation of the binding sites in these proteins.

A number of high affinity InsPn binding proteins have now been purified and identified (Table 1). Two major protocols have driven the rapid and efficient purification of these binding proteins. The first technique has taken advantage of the affinity of InsPn binding proteins for heparin–agarose or anionic resins as an early step in the purification. Heparin–agarose chromatography often yields a rapid, efficient (10- to 100-fold) purification in the first step. It has been proposed that rather than acting as a conventional cation exchange resin (since many of the proteins that interact strongly have neutral pIs), heparin–agarose, a polysulfated glycosaminoglycan, may act as a pseudo-affinity resin. After this initial step, either ion exchange, size exclusion, or InsPn affinity column chromatography has been used to isolate high affinity InsPn binding proteins. The second technique which has facilitated InsPn binding protein purification is the use of the polyethylene glycol (PEG) precipitation assay for measuring binding in solution. Previously, mini gel filtration column separations were used for binding assays for several solubilized receptors, including the InsP_3 receptor (95). But while these columns yield rapid and efficient separations, they suffer from the difficulty of assaying large numbers of samples, a task which is more easily accomplished with the PEG assay.

Of the ten high affinity InsPn/PtdInsPn binding proteins isolated (Table 1), four distinct protein complexes that are specifically involved in membrane vesicle trafficking have been identified. AP-2, a clathrin assembly/adaptor protein associated with clathrin coated vesicles at the plasma membrane (reviewed in ref. 96) was the first InsPn binding protein to be identified (97, 98) that bound InsP_4 and InsP_6 with high affinity. In the AP-2 heterotetramer, the α_A and α_C subunits represent the InsPn binding subunits. Clathrin binding activity, homo-oligomerization, and AP-2 stimulated clathrin cage formation *in vitro* are inhibited by InsP_4, InsP_6 (99), and

Table 1. Identified inositol polyphosphate and inositol phospholipid binding proteins

Identified protein	(Ptd)InsPn affinity	Reference(s)	Functional effect(s)
1. AP-2	$InsP_4$, K_d = 640 nM	97, 98	Inhibition of self-assembly, clathrin binding, and clathrin cage formation
	$InsP_6$, K_d = 120 nM	97–99	
	$PtdInsP_3$, K_d = 350 nM	100	
2. AP-3	$InsP_6$, K_d = 240 nM	102, 103	Inhibition of clathrin cage formation
	$PPInsP_5$, K_d = 20 nM	103	
	$PtdInsP_3$, IC_{50} = 5.8 μM	101	
3. Golgi coatomer	$InsP_4$, K_d = 0.1 nM	110, 111	Inhibition of coat formation and cation conductance
	$InsP_6$, K_d = 0.2 nM	110, 111	
4. Synaptotagmin II	$InsP_4$, K_d = 30 nM	115	?
	$PtdInsP_3$, K_d = 1 μM	154	
5. FKB12, erythrocyte immunophilon	$InsP_3$, K_d = 95.6 nM	126	Inhibition of peptidyl prolyl isomerase activity
6. GAP^{IP4BP}, platelet	$InsP_4$, K_d = 14.4 nM	126	Relieves inhibition of RAS GAP
7. 74 kDa, nuclear	$InsP_4$, K_d = 6.3 nM	125	?
8. Phospholipase Cδ1	$InsP_4$, K_d = 2.3 nM	127	?
	$InsP_3$, K_d = 3.4 nM	12	Facilitates membrane association and access to substrate
	$PtdInsP_2$, K_d = 1.8 μM	51	
9. Centaurin α	$PtdInsP_3$, IC_{50} = 120 nM	129	?
10. 42 kDa, cerebellar	$InsP_4$, K_d = 5.6 nM	120–122	?

PtdIns(3,4,5)P_3 (100). Another clathrin assembly/adaptor protein, AP-3, is a high affinity InsP_6 binding protein (101–103). While both AP-2 and AP-3 bind InsPns, AP-3 is composed of a single subunit, shares little sequence similarity to its functional relative AP-2 and is expressed exclusively in brain (104–106). Both bacterially expressed AP-3 and purified brain AP-3 bind InsP_6, PP-InsP_5, PtdInsP_3, and InsPn mediated inhibition of clathrin assembly occurs in direct proportion to binding affinity (101, 103). Studies with GST fusion proteins of AP-3 truncation mutants have localized the InsP_n binding domain to the amino terminal 40% of the AP-3 protein (102).

A third protein complex involved in regulation of vesicle trafficking, the Golgi coatomer complex (107) (see refs. 108 and 109 for a review), also exhibits high affinity for the highly phosphorylated InsPns, InsP_4 through PP-InsP_5 ('InsP_7'). Both bovine brain (110) and *S. cerevisiae* coatomer bind InsPns, with PP-InsP_5 > InsP_6 > InsP_5 (111). The affinities of the separate coatomer proteins for different InsPns have been recently examined by photo-affinity labelling with [^{3}H]BZDC-InsP_4 and -InsP_6 (Chaudhary and Prestwich, unpublished data and ref. 112). Surprisingly, only the α coatomer protein (α COP) was labelled by the InsP_4 photoprobe, while all the COPs labelled with the InsP_6 probe, suggesting that many coatomer subunits may contribute to binding, but with differing specificities.

A fourth component of a vesicle trafficking machinery, synaptotagmin, has high affinity for InsP_4, InsP_6, and PtdInsP_3. Synaptotagmin, a transmembrane protein that is localized to synaptic vesicle membranes (113) in neurons (reviewed in ref. 114), was purified as an InsP_4 binding protein (K_d = 30 nM) (115), but with slightly different specificity than AP-2 or AP-3, with InsP_5 > InsP_4 > InsP_6. By mutagenesis and photoaffinity labelling studies, the binding site for InsPns has been localized to the C2B domain (refs. 116 and 117, and B. Mehrotra, J. Elliot, and G. Prestwich, unpublished results) of the synaptotagmin I and II isoforms. This particular domain is involved in the functional regulation of vesicle release (117–119). The observation that InsPns bind to AP-2, AP-3, coatomer, and synaptotagmin suggests that PtdInsPn/InsPn binding may serve a ubiquitous role in regulating vesicle trafficking events. Moreover, the enrichment of these proteins in brain suggests a possible role for PtdInsPns/InsPns in modulation of neuronal activities, perhaps via regulation of synaptic vesicle release and/or recycling.

Four additional novel, high affinity InsPn binding proteins, which show selectivity for Ins(1,3,4,5)P_4 over other InsPns, have now been identified and are included in Table 1. First, a 42 kDa Ins(1,3,4,5)P_4 binding protein from porcine cerebellar membranes has been purified, which displays an acidic pH optimum (K_d = 5.6 nM) (120). By immunological analysis, this protein is found to be present in both membrane associated and soluble fractions from various neuronal and non-neuronal tissues (121, 122). Secondly, an Ins(1,3,4,5)P_4 binding protein, isolated from platelet plasma membranes, has been purified and the corresponding cDNA cloned (123). This purified

95 kDa Ins(1,3,4,5)P_4 binding protein has a high affinity for its ligand, with a $K_d = 6.2$ nM (124). Platelet p95 stimulates the GTPase activity of two small molecular weight GTP binding proteins, Rap and Ras, indicating that p95 has GAP activity, hence, the name of p95, GAP1^{IP4BP} (125). Ins(1,3,4,5)P_4 specifically regulates the Ras GAP activity. GAP1^{IP4BP} and the C2B domain of synaptotagmin I and II contain a positively charged amino acid motif that may coordinate the negatively charged phosphate groups on the inositol ring of Ins$P$ns.

Thirdly, a 12 kDa immunophilin (FKBP12) isolated from human erythrocyte membranes, a member of the rapamycin and FK506 binding protein class of peptidyl prolyl *cis-trans* isomerases, also binds Ins(1,4,5)P_3 and Ins(1,3,4,5)P_4 with high affinity (126). Interestingly, the prolyl isomerase activity of this FKBP12 is also strongly inhibited by nanomolar concentrations of Ins(1,4,5)P_3 and Ins(1,3,4,5)P_4. Finally, a nuclear 74 kDa high affinity Ins(1,3,4,5)P_4 receptor (IC$_{50}$ = 2.3 nM) has recently been purified from detergent solubilized rat liver membranes that exhibits an acidic pI of 4.3 (127). N-terminal peptide analysis shows that this nuclear Ins(1,3,4,5)P_4 receptor is a novel protein. Thus, in addition to roles in regulation of vesicle trafficking, Ins(1,3,4,5)P_4 may be involved in modulating other important physiological functions, including the Ras signalling pathway, protein folding, and nuclear calcium regulation.

2.2 PtdIns$P$n binding proteins

PtdIns(4,5)P_2 and PtdIns(3,4,5)P_3 contain inositol phosphate head groups that are structurally analogous to Ins(1,4,5)P_3 and Ins(1,3,4,5)P_4 respectively. Given these similarities, some of the techniques used to identify Ins$P$n binding proteins have proven to be useful for the identification of PtdIns$P$n binding proteins. For example, phospholipase C$_\delta$ contains a PH domain, distinct from its catalytic domain, which it is suggested binds PtdIns(4,5)P_2 *in vivo*. Phospholipase C$_\delta$ was originally isolated using an immobilized Ins(1,4,5)P_3 affinity column (12). A novel protein with homology to phospholipase C$_\delta$ has also been identified using an immobilized Ins(1,4,5)P_3 resin (128). In recent studies, photoaffinity labelling of the PH domain in intact phospholipase C$_\delta$ and the isolated PH domain was demonstrated using a [^{3}H]BZDC-Ins(1,4,5)P_3 photoprobe (51).

Recently, we have identified a 46 kDa protein from rat brain, which we have named centaurin α (129). This protein binds to an Ins(1,3,4,5)P_4 resin with high affinity, but does not bind [^{3}H]Ins(1,3,4,5)P_4 or the [^{125}I]ASA-Ins(1,3,4,5)P_4 photolabel (130, 131). We have now used the more hydrophobic and chemically more stable photoprobe, [^{3}H]BZDC-Ins(1,3,4,5)P_4 to label centaurin α (see section 6). The ligand with the highest affinity for centaurin α is PtdIns(3,4,5)P_3, (IC$_{50}$ = 120 nM), while Ins(1,3,4,5)P_4, PtdIns(4,5)P_2, and Ins(1,4,5)P_3 bind with 5-, 12-, and more than 50-fold lower affinity, respectively. The cDNA for centaurin α, cloned from a rat brain library, encodes a protein with a novel putative zinc finger domain and a number of ankyrin-like repeats, and which exhibits homology to a variety of proteins, including a liver Arf GAP and yeast growth regulatory proteins. Comparison of published partial amino acid sequence data from the 42 kDa cerebellar Ins(1,3,4,5)P_4 binding protein (122, 132) with centaurin α suggests that some domains may be highly conserved between the two proteins.

2.3 Summary

Many high affinity Ins$P$n and PtdIns$P$n binding proteins have now been identified. The data suggest that, in addition to the regulation of intracellular calcium, Ins$P$ns may play more ubiquitous roles as regulators of vesicle trafficking. However, neither the exact endogenous ligand nor the *in vivo* functional significance of PtdIns$P$n binding for these proteins has been resolved. Furthermore, the identification of several distinct proteins, including GAP1^{IP4BP}, an Ins(1,3,4,5)P_4-regulated immunophilin FKBP12, and a nuclear Ins(1,3,4,5)P_4 binding protein, demonstrate a potential for many points of action for just one Ins$P$n. This is turn suggests that there may be many more Ins$P$n targets involved in regulatory functions in the cell. No data have been obtained demonstrating a direct interaction of Ins(1,3,4,5)P_4 with a conventional ion channel, a channel associated protein, or calcium regulatory protein. This raises the possibilities that (a) these proteins have not yet been identified because of low abundance or instability; or (b) some of the modulatory calcium effects occur through less direct routes, perhaps via one of the identified proteins. Some indirect mechanisms of modulation may include the interaction of a channel complex with the cytoskeleton, the increased uptake of calcium by stimulated endocytosis, or the 'up-regulation' of one of the calcium homeostatic activities, e.g. via regulated recruitment of calcium channels to the plasma membrane from an intracellular compartment.

The data summarized demonstrate that identification of specific, high affinity InsP_4, InsP_6, and PtdIns(3,4,5)P_3 binding proteins holds significant potential for the elucidation of the function of these inositol-containing compounds in many tissues. Toward this end, some of the protocols describing techniques for ligand binding, purification of binding proteins, and photoaffinity labelling are further described in this chapter.

3. Theoretical considerations

3.1 Receptors and binding proteins

In order to begin to establish the functional roles of binding proteins for inositol lipids and inositol polyphosphates, it needs to be determined whether the binding protein is a classical second messenger receptor or a less conventional receptor. Many first and second messenger receptors transduce the binding of a ligand (whose levels increase transiently) into a change in the activity of the protein, such as modulation of enzymatic or ion channel activity or of the coupling of the receptor protein to downstream enzymatic/channel activities. $Ins(1,4,5)P_3$ fulfills the following 'classical' criteria for a second messenger. Its synthesis increases rapidly and transiently following ligand receptor interaction. It possesses a short half-life. There exist high affinity, high specificity binding proteins whose activities (channel gating) are modified by $Ins(1,4,5)P_3$; and exogenous addition of $Ins(1,4,5)P_3$ evokes calcium release from intracellular stores. Diacylglycerol fulfills a similar set of criteria as a messenger in activation of several PKC isoforms (see ref. 133). It has yet to be established, though, whether the other inositol derivatives similarly fulfill these classical second messenger criteria.

Even if the ligand concentration does not change during acute agonist stimulation however, binding may serve other critically important functions, such as transport, localization, regulation, or conformational switching. For example, binding of $PtdIns(4,5)P_2$ to the PH domain of the β-adrenergic receptor kinase has been proposed to regulate the interaction with the βγ subunits of GTP-binding (G-) proteins. In addition, $PtdIns(4,5)P_2$ binding has been proposed to serve as a mechanism by which specific proteins are recruited to specific membranes. $PtdIns(4,5)P_2$ binding to the PH domain of phospholipase C_δ has been proposed to allow the enzyme to bind to and to 'scoot' over regions of the membrane containing its substrate, $PtdIns(4,5)P_2$ (see ref. 134 for a discussion).

It is also possible that binding proteins act as molecular switches, analogous to G-proteins. These proteins are not sensors or receptors for GTP since GTP levels do not change dramatically during cell signalling. Rather, the binding of GTP (which is stimulated by receptors or GTP-exchangers) is conveyed into an activation signal, in the context of a molecular switch (135, 136). When GTP is hydrolysed to GDP, the G-protein is deactivated. If this is the case, then the roles of proteins which interact with putative $PtdInsPn$/$InsPn$ binding protein switches, as well as phosphatases, kinases, and proteins which modulate the affinity of $PtdInsPn$/$InsPn$ binding, are also worthy of examination.

Another potentially important role for $PtdInsPn$/$InsPn$ is the direct regulation of protein–protein interactions by the ligand, and, conversely, the modulation of the binding of $PtdInsPn$/$InsPn$ by other signals. Such regulation has been proposed for proteins such as AP-2, AP-3, coatomer, and synaptotagmin (103, 111, 137). It is possible that these proteins respond to changes in ligand levels via the regulation of their coupling to other proteins in oligomeric complexes. It is also conceivable that these proteins may be constitutively liganded under basal conditions. Then, during agonist stimulation, other signals generated could lead to ligand dissociation and changes in protein–protein coupling. Regulation of protein–protein interactions by small ligands such as PtdIns$P$ns and Ins$P$ns may be revealed to be an important mechanism in the regulation of protein complex assembly and disassembly.

3.2 Physiological levels of Ins$P$ns, and values for IC$_{50}$ and K_d

In a given cell, it is important to estimate the free mass levels of Ins$P$ns and how they change in response to agonist stimulation, so that predictions can be made regarding whether binding proteins persist in a ligand-bound or unliganded state and, in the former case, which ligand is most likely to occupy the binding protein—even though this approach may initially raise more questions than it answers. For example, the affinity of the Ins(1,4,5)P_3 receptor for its ligand *in vitro* is approximately an order of magnitude lower than the estimated free concentration of Ins(1,4,5)P_3 in resting cells. Possible ramifications of this observation include (a) that the Ins(1,4,5)P_3 receptor is always saturated by its ligand and is regulated by other factors, such as calcium, in acute signalling; (b) that the amount of Ins(1,4,5)P_3 that is 'accessible' to the receptor is substantially lower than the mass estimates, and thus the receptor is only liganded when receptor generated Ins(1,4,5)P_3 increases occur; or (c) that the affinity Ins(1,4,5)P_3 for its receptor *in vivo* is substantially lower than that determined *in vitro*.

These complications make it necessary to attempt to establish the free or accessible concentrations of Ins$P$ns, and whether Ins$P$ns are compartmentalized, sequestered, or complexed with metal ions such as calcium, magnesium and iron. Chelation of Ins$P$ns with metals, in particular calcium, has recently been proposed as a potential mechanism for acute regulation of the microdomain levels of Ins$P$ns (138). It is also important to consider the relative levels of different Ins$P$ns when considering which is the most likely ligand for any individual binding protein. For example, a site which binds InsP_4 and InsP_5 with nearly equal affinity may be physiologically most influenced by InsP_5, whose intracellular concentrations are at least 10 times greater than those of the other candidate ligand.

Radiolabelled and photoaffinity ligands provide probes that allow the characterization of the interaction between a ligand and its binding site. The affinity of a PtdIns$P$n/Ins$P$n binding protein for a particular ligand may be expressed as a K_d or IC$_{50}$ depending on whether the interaction is examined by direct or indirect methods. In direct binding assays, a measure of the direct interaction of a ligand with its receptor permits the determination of kinetic and equilibrium constants, as well as receptor density, which are used to optimize conditions for binding characterization. This gives the K_d which is an intrinsic property of the receptor, and is independent of ligand concentration used in the assay but requires a more detailed kinetic analysis to be undertaken and the availability of the ligand as a labelled analogue, such as a photolabel, or radiolabels such as [^{3}H] or [^{32}P].

In indirect binding assays, a pharmacological profile of a series of unlabelled ligands can be determined by measuring the inhibition of binding of a radioligand by these unlabelled compounds. The IC_{50} value generated by this method is the concentration of added unlabelled ligand which leads to a 50% reduction (or gives a 50% displacement) in the binding of the labelled ligand. The IC_{50} is an extrinsic property, a relative value, and depends on the concentration of labelled ligand used in the assay. While IC_{50} measurements are less informative when comparing data from different laboratories, measurements of IC_{50} values are still valuable since they allow comparison of the relative affinity/efficacy for a wide variety of compounds in their ability to displace the labelled ligand. Therefore, determination of IC_{50} is especially useful when a particular labelled ligand is limited by cost or availability. Furthermore, in photoaffinity labelling, which is a non-equilibrium binding, IC_{50} values can also be determined and compared with IC_{50} values obtained under equilibrium binding conditions. If the ligand concentration used in the assay is well below the K_d of the receptor, then the IC_{50} value can be close to the K_d. Finally, where a biological effect of adding an InsPn to a cell or system can be assessed, the relative ED_{50} (concentration required for half maximal response) can be determined. The ED_{50} value can then be compared with an IC_{50} value for different compounds to assist in identifying molecular targets and InsPns involved in particular biological responses.

4. Binding assays

The four most important factors to be considered in designing a binding assay are (a) how to separate the bound from the unbound ligand, (b) the specific activity of the ligand, (c) the density of binding sites, and (d) the affinity of the receptor and its dissociation rate. Of course, the latter two factors cannot be determined until after the assay has been performed. Use of a tissue/subcellular fraction with an enrichment of binding sites helps improve the signal (the ratio of total to non-specific binding), especially if low specific activity ($[^3H]$-labelled) ligands are employed. The speed with which the bound ligand is separated from the free, and the efficiency of this separation are both important parameters. Most binding reactions do not exhibit a high temperature dependence, and thus are performed at $0°C$ so that metabolism of the ligand and protein during the binding reaction is minimized. In addition, most receptors have fairly fast association rates that are usually diffusion limited, and binding normally equilibrates within approximately 5–30 min. The factor that usually contributes most to the K_d is thus the rate of dissociation of the ligand. This rate is very high for low affinity (high K_d) receptors (half times in the order of seconds) and substantially slower for high affinity (low K_d) receptors (half times in the order of hours). Most of the InsPn binding proteins identified have K_d values in the nanomolar range with corresponding dissociation half times between many minutes and several seconds. Therefore, rapid separation time is essential in separating bound from free ligand, and a 'wash' step, which can often reduce the non-specific binding and improve the signal-to-noise ratio, is usually inappropriate, since the ligand dissociates within the time it takes to perform the wash.

4.1 Protocols for Ins*P*n binding proteins

4.1.1 Membrane binding

In membrane fractions, the separation of bound from unbound ligand is straightforward since membranes suspended in low osmotic strength buffers are easily sedimented by microcentrifugation, which allows for a very rapid separation of the membrane bound from the free ligand in the absence of dilution steps. Moreover, numerous samples can be assayed in a short time so that various parameters of binding can be characterized. Membranes can first be fractionated by differential or sucrose gradient sedimentation, or with an electric field, and then used in a microcentrifugation binding assay. In measurements of InsP_6 binding in crude samples, the 'unmasking' of binding sites during purification (131) suggests that, in the crude fraction, endogenous inhibitors of InsP_6 binding may be present. Thus, it may be advisable to wash membranes before binding if not using fractionated membranes. In these purified membrane preparations, binding studies require the use of high specific activity radiolabelled Ins*P*n probes since the receptor levels (B$_{max}$) are often low.

Two other methods for measurement of binding in membrane fractions are worthy of mention, namely filtration and blot overlay. In filtration, separation of bound and free ligand is performed through a filter apparatus such as a 'Brandel' or a 'Millipore' manifold, a method which has been used extensively to identify and characterize traditional hormone, neurotransmitter, and growth factor receptors. These conventional filtration assays use glass fibre filters. Such filters can only be used for trapping membranes and can bind the Ins*P*n ligand (both bound and free). Thus, they must be washed to reduce the filter binding. A blot overlay assay, which is not described in this chapter, has been used to identify several InsP_3 binding proteins (139). This type of assay has also been used successfully to identify calcium, calmodulin, and GTP-binding proteins. The nitrocellulose blot is incubated with [^{32}P]Ins(1,4,5)P_3 and then subsequently washed, dried, and exposed to film to reveal several protein bands which have bound the [^{32}P]- labelled probe. Using radio-labelled Ins(1,4,5)P_3, eight polypeptides which migrated on SDS gels between 23 and 50 kDa were detected in separated rat liver membranes.

Ins(1,3,4,5)P_4 binding to crude membranes from numerous tissues has been reported (see section 1 for references). Usually the amount of specific binding is directly proportional to the amount of binding protein added in the assay. Thus, increasing the amount of membranes added usually increases the amount of specific binding but does not affect the ratio of total to non-specific binding. The binding is also directly proportional to the concentration of ligand used in the assay, until saturation (at a concentration of approximately twice the K_d) is reached. Again, increasing the concentration of ligand does not improve the signal, but rather increases the total amount of specific binding, yielding more reliable measurements. The amount of ligand used in the assay is usually limited by the cost and availability. The K_d is often determined either by doing a 'hot Scatchard', i.e. by increasing the concentration of radiolabelled ligand in the assay. For high affinity receptors (K_d < 100 nM) this is practical, but for lower affinity receptors, this is often not feasible, given the cost of the ligand. In these cases, a 'cold Scatchard' is performed, by maintaining a constant concentration of labelled ligand and increasing the concentration of unlabelled ligand, with a corresponding dilution of specific activity of the ligand. Below the K_d (where binding is linear with ligand concentration), as the unlabelled concentration increases, there is no decrease in total binding, because the amount of binding increases proportionally to the decrease in specific activity. Once the binding saturates, the dilution of specific activity is not paralleled by an increase in binding and there is a substantial decrease in the measured radioactivity. Non-specific binding should be performed with a 100-fold excess of unlabelled ligand.

Protocol 1. Reversible binding of radiolabelled Ins*P*ns in membrane fractions

Equipment and reagents

- Tools for dissection
- Polytron (e.g. Kinematica, Brinkman, or Cole-Parmer) or Motorized teflon-glass tissue homogenizer (e.g. Glas-Col or Cole-Parmer)
- Super-speed centrifuge (30 000 × g) and 50 ml plastic round bottom tubes (Beckman)
- Microcentrifuge and 1.5 ml microcentrifuge tubes
- Scintillation counter, vials, and cocktail
- Radiolabelled ligand, i.e. [³H]Ins(1,3,4,5)P_4, [³H]InsP_6 (Amersham, DuPont Co.); see ref. 83 for synthesis protocol for [³²P]Ins(1,3,4,5)P_4)
- Unlabelled ligand prepared at 10 × concentration (Calbiochem, Sigma)

- Homogenization buffer (HB): 50 mM Tris-HCl pH 7.7, 1 mM EDTA, 1 mM EGTA, 1 mM 2-mercaptoethanol, containing the following protease inhibitors at the stated final concentrations: 100 μg ml⁻¹ phenylmethanesulfonyl fluoride (PMSF), 250 μg ml⁻¹ N-carbobenzoxyphenyl-alanine (CBZ-Phe), 5 μg ml⁻¹ antipain, 10 μg ml⁻¹ leupeptin, and 10 μg ml⁻¹ aprotinin, 5 μg ml⁻¹ chymostatin, and 5 μg ml⁻¹ pepstatin A. All proteases are prepared in H₂0, except for PMSF and CBZ-Phe which are dissolved in 100% ethanol or methanol, and chymostatin and pepstatin A in DMSO
- Binding buffer (BB): 25 mM Tris-HCl pH 7.7, 1 mM EDTA
- 1% Sodium dodecyl sulfate (SDS) in H₂O

A. *Preparation of crude membranes from whole brain*

1. Prepare HB, chill to 4°C, and add protease inhibitors from frozen stocks thawed immediately before dissection, except for PMSF and CBZ-Phe which should be made fresh.

2. Dissect tissue as rapidly as possible after sacrifice. Place one whole dissected rat brain (100–300 g male Sprague-Dawley rat, 1.0–1.3 g

wet weight) into 15 volumes (20 ml) HB in a 50 ml centrifuge tube. Previously frozen brains, thawed on ice, can also be used.

3. Gently homogenize the tissue (Polytron setting 6–7 for 7–15 sec or a Teflon-glass homogenizer with 10 strokes) at 0 °C.

4. Pellet total crude membranes by centrifugation for 10–15 min at 30 000 × g at 4 °C.

5. Wash the membrane pellet by resuspending and rehomogenizing in 20 ml HB and centrifuging again as in step 4. Membranes can be washed a second time if necessary. Then resuspend in a final volume of 15 ml HB without protease inhibitors. The protein concentration is between 5–10 mg ml^{-1}.

B. *Centrifugation Binding Assay*

1. In a 1.5 ml microcentrifuge tube, incubate 30–100 μl of membranes (0.3–1.0 mg protein) with 0.001–0.100 μCi of radiolabelled ligand in a final volume of 100–400 μl BB for 10–30 min on ice. All conditions should be assayed in duplicate or triplicate.

2. Separate the bound ligand from free by microcentrifugation for 3–10 min at 13 000 r.p.m. (16 000 × g) at 4 °C.

3. Aspirate the supernatant using a 200 μl Pipetman yellow pipet tip. Pellets are usually quite tight and the excess supernatant can be removed by placing the tip close to the pellet. While our laboratory does not routinely do so, some groups perform a brief rinse of the pellet with ice-cold H$_2$0 after centrifugation; for an example, see ref. 85.

4. Solubilize pellets by adding 100 μl of 1% SDS for at least 1 h or overnight at room temperature. Alternatively pellets can be solubilized in 0.5 N NaOH.

5. Transfer the solubilized pellet to a scintillation vial; add 5–7 ml scintillation cocktail.

6. The assay should be designed to vary the concentration of the ligand over many orders of magnitude, typically from as low as 0.1 nM to 10 μM. The sample, containing only labelled ligand, is designated as the 'total' binding (performed with 0.2–20 nM labelled ligand, depending on how many μCi are added and the final volume). The 'non-specific' binding is determined in a separate tube by the addition of between 100- to 1000-fold excess concentration of unlabelled ligand. The 'specific' binding is determined by subtracting the 'non-specific' binding from the 'total' binding.

C. *Alternative to microcentrifugation: membrane filtration*

1. Incubate membranes with labelled ligand as described in step 1 of *Protocol 1B* above in a final volume of 0.5 ml.

Protocol 1. *Continued*

2. Pass the binding mixture directly through a glass-fibre filter (Whatman GF/C) under vacuum. Alternatively, the binding mixture can be diluted in 3 ml of 'dilution buffer'—ice-cold phosphate buffered saline (pH 7.0) containing 1 mM $InsP_6$—and then filtered.

3. Wash the filter with 2 ml of binding buffer or twice with 3 ml of the dilution buffer in step 2, allow to air dry and count for radioactivity. See refs. 77 and 140 for data generated using this type of separation.

Notes
1. In a typical centrifugation assay with 0.6 mg of rat brain membranes and 10 nM labelled $Ins(1,3,4,5)P_4$, the total binding in brain is approximately 1600 c.p.m. and the non-specific binding about 400 c.p.m., or a ratio of about 4:1. This is not an ideal binding ratio. In a binding assay, a 10:1 ratio would be preferred. Because the pellets cannot be resuspended and washed to remove the trapped ligand, which contributes to the non-specific binding, this is often impossible to improve.
2. When converted to specific activity of binding, this yields a binding of 1 pmol mg^{-1} protein. Given a K_d of 100 nM, this yields a B_{max} of approximately 10 pmol mg-1 protein.
3. Parameters which can be varied and adapted to a particular assay include: concentration of labelled ligand, tissue source and amount of membranes added in the assay, ionic strength, pH, metal ion (such as calcium or magnesium) concentration, and inorganic phosphate addition. Finally, to use enriched membrane fractions, fractionation of membranes can be performed. Recommended protocols for membrane fractionation: for liver fractionation, see Malviya et al. (141), for platelets fractionated by free-flow electrophoresis, see Crawford et al. (142), and for brain membrane fractionation see Gordon-Weeks (143).
4. The use of other filters for the filtration assay could be explored. For example, nitrocellulose membranes (protran BA 85, 0.45 mm, Schleicher and Schuell) could potentially be used.

4.1.2 Autoradiography

Autoradiography allows for the regional and cellular distribution of binding proteins to be determined. Originally used for cell surface receptors such as those for dopamine, this technique has now been applied to intracellular binding proteins and enzymes as well (144). The advantage of autoradiography is that it allows for the identification of specific populations of cells that express binding proteins, even when those populations are only a small percentage of the total tissue. It can (a) identify a useful source of tissue for purification, (b) identify exactly which cells in a population express the binding proteins, (c) allow for comparisons between tissues and species, (d) permit a detailed analysis of developmental regulation, and of changes which accompany disease or trauma (such as lesions), and (e) can be especially useful for the characterization of a new ligand. In this technique, a thin layer of binding assay mixture is layered onto a thin (10–30 μm) tissue section. The tissue is not fixed as it is for immunohistochemical or *in situ* localization, but is sectioned from unfixed frozen blocks. Only the top few layers of cells are accessible to the ligand. Including a low concentration of detergent such as digitonin can help to permeabilize more cells and expose more receptors. The section is then washed very briefly (so that the specifically bound ligand does not dissociate), dried, and exposed to film for lower resolution or emulsion coated coverslips for higher resolution analysis.

Protocol 2. Autoradiographic localizations with radiolabelled Ins*P*ns

Equipment and Reagents

- Cryostat microtome (e.g. Jung or IEC)
- Tissue-tek (embedding medium/OCT compound from Miles Laboratory) and microtome chucks
- Autoradiographic film for detection of [³H], such as [³H]Hyperfilm (Amersham)
- Perfusion buffer (PB): 50 mM NaHPO₄ pH 7.5, 0.3 M Sucrose

- Glass microscope slides
- NTB-2 or NTB-3 emulsion (Kodak), Dektol or D19 developer, and fixer (Kodak) for light microscope autoradiography
- Autoradiography buffer (AB): 20 mM NaCl, 20 mM Tris pH 7.7, 100 mM KCl, 1 mM EDTA). The pH of the buffer can be modified for optimization of binding.

A. Preparation of sections

1. Anesthetize one male Sprague-Dawley rat with pentobarbital (60 mg kg⁻¹ body weight).

2. Perfuse the rat via the left cardiac ventricle with PB and sacrifice.

3. Remove the brain and freeze it in Tissue-tek on a microtome chuck. Alternatively, the brain can be immersed for 15 sec in 2-methylbutane at −40°C and then kept frozen at −80°C until sectioning.

4. Section brain tissue using a cryostat, cutting 10–15 μm sections and thaw-mounting onto subbed slides (0.5% gelatin) or polylysine (0.1% w/v) coated slides. Sections on slides are stored at −20°C or −80°C until use.

B. Binding incubation

1. Pipet 50–100 μl AB containing 10–50 nM (final concentration) [³H]- or [³²P]-labelled ligand onto the section to form a 'puddle' that covers the section and incubate for 10–15 min at 4°C.

2. Wash the slides twice for 0.5–2.0 min in ice-cold AB or 25 mM sodium phosphate buffer of the appropriate pH.

3. Dry the slides under a cool air stream and expose the slides to film or coat with NBT-2 or NBT-3 emulsion and expose for 2–6 weeks, depending on the specific activity of the Ins*P*n and the density of binding sites. When developing autoradiographic emulsions, Kodak Dektol or D-19 developer, and Kodak fixer are recommended. Instructions for dipping and development are included with the emulsion.

4. Non-specific binding is determined in parallel sections by including 1–10 μM unlabelled ligand. The approximate affinity can also be determined, as with membrane binding, by including increasing concentrations of unlabelled ligand.

Protocol 2. *Continued*

5. Autoradiographic densities can be quantified using a computerized microdensitometer or by light level microscopic analysis.

Notes

(a) When the pH dependency was determined for other Ins*P*n binding proteins in addition to the Ins(1,4,5)*P*₃ receptor (84), sections were preincubated for 1 h at 4°C in buffer with 25 mM sodium phosphate (appropriate pH), 20 mM NaCl, 100 mM KCl, and 1 mM EDTA. For binding, a 25 mM sodium phosphate buffer was used with 0.1% BSA, and 5 nM labelled ligand and binding was equilibrated for 30 min, and washed and dried as described.

(b) 1 mg ml⁻¹ BSA can be included in the binding to reduce the level of non-specific binding. Addition of 5 mM digitonin can help improve the binding signal since it permeabilizes the cells and allows for greater access of the ligand.

4.1.3 Reversible Ins*P*n binding in soluble, solubilized, or column fractionated samples

The binding assay used most frequently for Ins*P*n binding to soluble, detergent solubilized membrane fractions and purified fractions is the polyethylene glycol (PEG) precipitation assay. This assay was used in the identification of AP-2, AP-3, synaptotagmin, coatomer, p95 GAP^{IP4BP}, and the p74 nuclear Ins(1,3,4,5)P_4 binding protein. The binding is allowed to pre-equilibrate and then the free Ins*P*n is separated from the bound Ins*P*n by precipitating the total protein mixture with PEG. Both soluble and detergent solubilized proteins can be precipitated with their bound ligands by addition of PEG, which apparently does not interfere with the binding. It is not clear whether the ligand re-equilibrates after addition of PEG during the precipitation time. In purified samples, addition of carrier protein, either gamma globulins or serum albumin, helps to precipitate the protein efficiently. The non-specific binding is quite low since the pellets are relatively small and only a small amount of ligand becomes trapped in them.

Two other techniques to measure binding in solution are membrane filtration assays and 'spun' column assays. The spun column assay is also described here because of its potential use in PtdIns*P*n binding assays. However, to our knowledge, this assay has been used only for the Ins(1,4,5)P_3 receptor and a few Ins(1,3,4,5)P_4 binding proteins. Thus, other Ins*P*n ligands must be tested to determine how they will behave on a size exclusion column and to choose the optimal exclusion resin for each ligand and binding protein being tested. The bound ligand usually exhibits a molecular mass of greater than 20 kDa, and is thus separated from the free ligand (usually less than 2 kDa) by the molecular mass difference. Usually a gel filtration/size exclusion resin with an exclusion limit of 5–20 kDa (such as a Bio-rad P-4 or P-10) is used. Both soluble and detergent solubilized fractions can be assayed with this technique. Brief centrifugation of the samples is recommended to increase the speed of column separation, ensuring that dilution and dissociation does not occur during the column separation.

Protocol 3. PEG precipitation assay for measuring Ins*P*n binding in solution

Equipment and reagents

- Microcentrifuge and 1.5 ml microcentrifuge tubes
- Scintillation counter, vials, and cocktail
- Fraction containing binding proteins (crude soluble, detergent solubilized membranes, partially purified, recombinant, or purified protein)
- Radiolabelled ligand
- 25% Polyethylene glycol 3350 (PEG)

- Unlabelled ligand (prepared at 10 × concentration)
- Binding buffer (BB): 25 mM Tris-HCl pH 7.4, 1 mM EDTA, 1 mM EGTA, and 1.0 mM K_2HPO_4
- 5 mg ml^{-1} Bovine gamma globulin (BγG)
- 5 mg ml^{-1} Bovine serum albumin (BSA)
- 1% Sodium dodecyl sulfate (SDS) in H_2O

Method

1. In a microcentrifuge tube, incubate 5–20 μl of each fraction with 0.001–0.100 μCi of radiolabelled ligand to a final volume of 50–200 μl BB for 10–20 min on ice.

2. Add 50 μl of ice-cold BγG, 50 μl BSA, and 1 ml of 25% PEG, vortex, and incubate for 10–30 min on ice.

3. Separate the bound ligand from free by microcentrifugation for 5–10 min at 13 000 r.p.m. (16 000 × *g*).

4. Aspirate the supernatant into an appropriate radioactive waste container.

5. Solubilize the pellet by adding 100 μl of 1% SDS for at least 1 h or overnight at room temperature and vortex.

6. Transfer solubilized pellet to a scintillation vial. Add 5–7 ml scintillation cocktail.

7. Non-specific binding is determined in a separate reaction by the addition of unlabelled 3–10 μM ligand. Determine specific binding by subtracting non-specific counts from total counts.

Notes
Possible variations to the method include:
(a) Use fractionated soluble or detergent solubilized membranes, purified samples, or recombinant proteins.
(b) Modify pH, metal ion concentration, and/or phosphate ion concentration.
(c) Add low concentrations of detergents (0.1% CHAPS or 0.01% Triton X-100 final concentration) to help precipitate the proteins, especially in soluble fractions.
Alternative PEG assay adapted from Chadwick et al. (145):
1. Incubate the protein fraction in 50 μl final volume containing 25 mM Tris pH 7.5, 5 mg ml^{-1} BγG (or equine immunoglobulin), 100 mM KCl, 1 mM EDTA, 1 mM DTT, 10 nM (20 000) c.p.m. [^{3}H]Ins*P*n.
2. After 20 min, precipitate the protein by addition of 35 μl 30% (w/v) PEG (final PEG concentration 12%), followed by vortexing, and further incubation for 10 min on ice.
3. Centrifuge for 10 min at 10 000 × *g* at 4°C.
4. Aspirate the supernatants and wash the pellets with 1 ml of ice-cold incubation buffer minus BγG.
5. Dissolve the pellets in 1 ml of 1% SDS, transfer to scintillation vials, add scintillation fluid, and determine radioactivity.

Protocol 4. Spun columns as an alternative assay for determining binding in solution

Equipment and reagents

- Centrifuge that can accommodate 3–8 cm columns (a clinical table-top centrifuge with swinging bucket rotors is recommended)
- Size exclusion/gel filtration resin that will include the unbound ligand (Biogel P-4 or P-10 resins (polyacrylamide beads), Biorad) and exclude the protein-bound ligand
- Scintillation counter, vials, and cocktail
- Unlabelled ligand (prepared at 10 × concentration)
- Spun columns: S/P pipetter tips for Selectapette Pipetter (Becton Dickinson Labware/Clay Adams) with glass beads, or disposable minicolumns (1 ml) with frits.
- Fraction containing binding proteins (crude soluble, detergent solubilized membranes, partially purified, recombinant, or purified protein)
- Radiolabelled ligand
- Binding buffer (BB): 25 mM Tris-HCl pH 7.4, 1 mM EDTA, 1 mM EGTA

Method

1. Swell the gel filtration/size exclusion resin in H_2O for at least 24 h before using.

2. Pipet the resin into the columns (1 ml packed resin). Allow excess buffer to run through the column so that the top of the resin is exposed.

3. Incubate the sample in BB in a total volume of 200 μl with 0.1–20 nM labelled Ins*P*n. For detergent solubilized samples, the detergent must be maintained in the BB. Triton X-100 or Brij 58 at 0.1% final concentration is recommended. Equilibrate binding for 10–30 min on ice.

4. Gently layer the binding reaction on the column and immediately spin at 1000 × *g* for 6 min or 3500 × *g* for 2 min. Columns are spun directly into scintillation vials to collect the void volume, or the sample can also be spun into a 15 ml plastic conical tube and then transferred to a scintillation vial.

5. Add the scintillation cocktail and count.

6. To determine non-specific binding, 1–10 μM unlabelled Ins*P*n is included in the assay. To determine the affinity and specificity of the binding protein, increasing concentrations of Ins*P*ns are included with the labelled ligand in the binding reaction, as described in *Protocols 1–3*.

Notes:
(a) When using samples with low protein concentrations, 1 mg ml^{-1} BSA can be included in the binding assay to improve recovery and separation.
(b) The development of a filtration assay to measure binding in solution could also be explored. Glass fibre filters do not retain soluble proteins and they cannot be coated with polyethyleneamine (PEI) for an Ins*P*n binding protein assay since the PEI binds the ligand. Nitrocellulose membranes (protran BA 85 0.45 mm, Schleicher and Schuell), which are used extensively for assays of GTP binding to G-proteins, may also be tested.

5. Purification of PtdInsPn/InsPn binding proteins

5.1 Heparin–agarose chromatography

As discussed previously, the use of heparin–agarose has been an important tool which has advanced the identification of InsPn binding proteins. It was first used for purification of the Ins(1,4,5)P_3 receptor, after it was discovered that heparin is a potent competitor of Ins(1,4,5)P_3 binding (95) and an antagonist of intracellular Ins(1,4,5)P_3-stimulated calcium release (146). Heparin–agarose chromatography has several advantages. It is fairly inexpensive, especially given the fact that it can be recycled and re-used. It has reasonably high capacity and yields consistent separations. We have found that by including 250 mM NaCl in the initial loading buffer, we can substantially enhance the capacity of the resin (by reducing the level of lower affinity binding). Thus, this resin can be used in the early steps of purification; we recommend a ratio of approximately 1 ml of resin per 30–100 mg protein. Both soluble and detergent solubilized proteins can be purified with heparin–agarose, though with the soluble fraction, we recommend a large dilution of the supernatant (3–7 fold) prior to incubation with heparin–agarose, because of the potential presence of factors such as endogenous InsPns which may interfere with binding to the heparin–agarose. One protocol described uses a batch incubation technique, which is fast and efficient, though several groups have also used gradient elution very successfully. In another protocol, prepacked heparin–agarose columns (which can also be recycled and re-used) can be adapted to the FPLC. Most of the InsPn binding proteins identified elute from heparin–agarose at between 0.4 M to 1.5 M NaCl.

Protocol 5. Heparin–agarose chromatography

Equipment and reagents

- 10–20 Sprague-Dawley male rats (150–300 g)
- Heparin (type I)–agarose resin (Sigma H6508)
- Kontes or Bio-rad glass columns (50 or 100 ml capacity)
- Rotator that can be placed in chromatography refrigerator or cold room
- Centriprep-30 concentrators (Amicon)
- Homogenization buffer (HB): see *Protocol 1*

- Supernatant dilution buffer (HB without protease inhibitors)
- Heparin–agarose wash buffer (HB containing 250 mM NaCl)
- Heparin–agarose elute buffer (HB containing 1.0–1.5 M NaCl)
- 20% 3-[(3-chloamidopropyl)dimethyl-ammonio]-1-propane-sulfonate (CHAPS)
- 5.0 M NaCl in H_2O

A. *Preparation of crude supernatant and detergent solubilized membranes*

1. Freshly prepare HB and chill to 4°C.

2. Dissect tissue as rapidly as possible after sacrifice. Place each dissected whole rat brain (1.0–1.3 g wet weight) into 20 ml HB in a centrifuge tube.

Protocol 5. *Continued*

3. Homogenize tissue (Polytron setting 6 for 8 sec or 10 strokes with a Teflon-glass homogenizer) at 0 °C in HB.

4. Separate membranes and cytosolic (crude supernatant) fraction by centrifugation for 15 min at $30\,000 \times g$ at 4 °C. Decant the supernatant into a clean 750 ml tissue culture flask.

5. Resuspend the membrane pellet by homogenizing in 20 ml HB containing 250 mM NaCl. Then add (1.1 ml) 20% CHAPS (1% final concentration). Invert tube several times to mix.

6. Solubilize the membranes on ice for 45 min.

7. Separate the solubilized membrane proteins from the insoluble fraction by centrifugation for 30 min at $30\,000 \times g$ at 4 °C.

B. *Batch heparin–agarose chromatography*

1. Wash heparin–agarose resin with at least 20 volumes of 0.5 M NaCl in H_2O. Approximately 1.0 ml of packed resin per brain is used for each of the supernatant and the solubilized membrane fractions. For example, 10 ml of packed resin is used for incubating the supernatant from 10 rat brains.

2. Dilute the crude supernatant fraction between 3- and 6-fold with ice-cold supernatant dilution buffer in a 750 ml tissue culture flask. Add 5.0 M NaCl to adjust the final concentration to 0.25 M. Then add the washed heparin–agarose resin.

3. Incubate the resin for a minimum of 1 h (but do not exceed 10 h) on a rotator at 4 °C.

4. Incubate the detergent solubilized membrane fraction with heparin–agarose in a flask as with the supernatant, using 1 ml of packed resin per brain, at 4 °C on a rotator. However, do not dilute the solubilized membrane proteins before incubation. Note that the NaCl concentration is already 0.25 M, NaCl having been added when the membranes were solubilized, so no additional NaCl is required before adding the resin.

5. After incubation, collect the heparin–agarose resin in a column by passing the resin and supernatant over a column.

6. Wash the resin with at least 15 volumes of HB containing 250 mM NaCl. For the solubilized membrane proteins the wash should also contain 1% CHAPS.

7. Elute the resin in the column by adding 7.5 volumes of HB containing 1–1.5 M NaCl for a minimum of 1 h (maximum overnight) on the rotator. For a 10 brain preparation/10 ml resin, this would require 75 ml of heparin–agarose elute buffer. The solubilized membrane heparin–

agarose elute buffer should also contain 1% CHAPS. Collect the eluate from the column.

8. Concentrate the eluate using Centriprep-30 concentrators (Amicon). The NaCl can also be reduced at this step by dilution and reconcentration of the eluate.

Notes
1. Triton X-100 has also been used successfully for solubilization of several Ins*P*n binding proteins. We recommend solubilization with 1.0% Triton X-100, and then reduction of the detergent concentration to 0.1% in the heparin–agarose wash and eluate buffers.
2. We routinely use the heparin–agarose resin at least three times. We recycle by washing the heparin–agarose with 20 volumes each of 2 M NaCl, 0.1% SDS, 2 M NaCl, and H$_2$0 in succession.
3. Cullen et al., (124) have also used prepacked heparin–agarose columns successfully to purify the GAP1^{IP4BP}. Prepacked heparin-columns (Hi-Trap) can be purchased from Pharmacia or Sigma, and are easily adapted to the HPLC. After an intial purification step using, for example, CM-Cellulose (124) the peak of binding activity is diluted and loaded on the Hi Trap column (5 ml) and then a gradient of 0.2M to 1.0M NaCl is used to elute the Ins(1,3,4,5)P_4-binding protein. A substantial purification is achieved in this step.

5.2 Ins*P*n affinity chromatography

Several groups have now used immobilized Ins*P*ns for the purification of binding proteins. Hirata and colleagues have used a C-2 hydroxyl-linked Ins(1,4,5)P_3 to purify phospholipase C_8 and a novel 130 kDa phospholipase C_8^--related protein (12, 128). Prestwich and coworkers have synthesized a P-1-tethered Ins(1,4,5)P_3 resin for the purification of Ins(1,4,5)P_3 receptors (147). Using a P-1–aminoethyl Ins(1,3,4,5)P_4 probe, Reiser and colleagues have purified and labelled a novel 42 kDa protein from porcine cerebellar membranes (148).

Researchers in our group have used a P-1–3-aminopropyl Ins(1,3,4,5)P_4 affinity resin to identify several high affinity Ins*P*n binding proteins from detergent solubilized rat cerebellar membranes with apparent molecular masses of 115/105, 182, 174, 84, and 46 kDa (130, 131). Confirmation of the ability of these proteins (except the 46 kDa protein) to bind Ins*P*ns was determined with a reversible [^{3}H]Ins(1,3,4,5)P_4 binding assay and by photo-affinity labelling with a radioiodinated azidosalicylamide derivative of Ins(1,3,4,5)P_4 ([^{125}I]ASA-Ins(1,3,4,5)P_4). The identified proteins displayed substantial differences in affinity, specificity, regional localization within the brain, and effects of pH, magnesium and calcium on binding, suggesting that they are separate Ins*P*n binding proteins. Further studies have revealed the identity of two of the five Ins*P*n binding proteins: the 115/105 kDa protein is AP-2 and the 46 kDa protein is centaurin α. The identities of the 182, 174, and 84 kDa proteins are still under investigation. The affinity resins mentioned here are also listed in the Appendix. As of this writing, these materials are not commercially available. Investigators interested in establishing a collaborative project involving these (or other customized) probes should contact Glenn D. Prestwick at his present address.

The binding proteins can be eluted from Ins*Pn* affinity resins with the appropriate concentration of Ins*Pn*. However, because of the cost of Ins(1,3,4,5)P_4, we found that it to be more economical to elute the proteins with increasing ionic strength buffers. This also allowed for an efficient wash and removal of the lower affinity binding proteins, as well as an efficient separation of the various Ins*Pn* binding proteins by an increased NaCl gradient. Elution of a Mono S column with InsP_6 also led to the identification of AP-3 as a high affinity Ins*Pn* binding protein.

Protocol 6. Ins*Pn* affinity chromatography

Equipment and Reagents

- Rotator that can be placed in chromatography refrigerator or cold room
- Centriprep-30 concentrators (Amicon)
- Kontes or Bio-rad glass column (50 or 100 ml capacity)
- Affinity column buffer (ACB): 50 mM Tris-HCl pH 7.4, containing 1 mM EDTA
- Affinity column elution buffer (ACEB): 200 ml ACB with dissolved 23.2 g NaCl (2 M final)
- 5 M NaCl
- 20% CHAPS

- Partially purified Ins*Pn* binding proteins (heparin–agarose column eluate, *Protocol 5*)
- Ins(1,3,4,5)P_4 affinity resin (synthesized by Prestwich and coworkers)
- FPLC (Pharmacia with 10 cm × 3 cm column for 10 ml of Ins(1,3,4,5)P_4 affinity resin)
- 0.45 μm filter bottles

A. *Batch purification*

1. Heparin–agarose column eluate concentrated 10-fold and diluted to the original volume with ACB to lower the NaCl concentration. All buffers for solubilized membrane proteins should contain 1% CHAPS.

2. Incubate the concentrated diluted eluate with resin for 2 h on a rotator at 4°C. Do not exceed 10 h.

3. Collect the resin in a column by passing the resin and the diluted eluate over a column.

4. Wash the resin with at least 10 volumes of ACB containing 100 mM NaCl.

5. Elute the adherent proteins from the resin with ACEB for a minimum of 1 h on the rotator. Collect the eluate from the column.

B. *Fractionation*

1. Filter all buffers using the 0.45 μm filter bottles. Buffer A will be ACB and buffer B will be ACEB. Remember to add CHAPS if purifying solubilized membrane proteins.

2. Set the FPLC for a flow rate of 0.2 ml min^{-1}.

3. Load the concentrated-diluted heparin–agarose column eluate on the Ins(1,3,4,5)P_4 affinity column at 5% buffer B.

4. After loading, wash the column at 5% buffer B for an additional 10 ml.
5. Increase the proportion of buffer B from 5% to 100% over 50 ml.
6. Collect 2 ml fractions and assay for binding, photolabelling, and/or protein composition.

6. Photoaffinity labelling

There are several major advantages to employing photoaffinity labelling, especially in parallel with reversible binding and/or affinity chromatography. The advantages have recently been enumerated for benzophenone-Ins*P*n probes (149). Two major advantages are (a) that it allows the identification of Ins*P*n binding proteins in crude and partially purified samples and (b) that it allows the determination of the protein subunit(s) that contain(s) the binding site(s). Because photoaffinity labelling produces a covalent attachment, labelled proteins can be denatured and resolved by SDS-PAGE, followed by autoradiography, thereby allowing the identification of binding subunits in crude fractions and in oligomeric protein complexes. This can be useful since purification may require considerably more material and be considerably more time consuming. Combined with affinity chromatography, photoaffinity labelling has proven itself a powerful tool for the identification and characterization of binding proteins. In the future, photoaffinity labelling will provide significant technical advantages for mapping binding sites.

Appropriate consideration must be given to the design of photoaffinity ligands. For the Ins*P*n probes that we have used, assumptions were made and tested about how derivatization of the $Ins(1,3,4,5)P_4$ ligand would affect its binding. Based on the observation that the $Ins(1,4,5)P_3$ receptor tolerated modifications at the P-1 position, we chose to synthesize $Ins(1,3,4,5)P_4$ probes modified at this position as well. This was a fruitful strategy since four binding proteins were isolated and identified with both P-1 tethered resin and photoaffinity probes. From the specificity measurements, it is likely that the 3,4, and 5 positions on the inositol ring are the most critical in determining binding for the proteins identified. Hence, these identified Ins*P*n binding proteins may not tolerate modification at positions other than P-1. In contrast, other potential $Ins(1,3,4,5)P_4$ binding proteins may not have been detected with these P-1 probes, if they cannot tolerate P-1 modification. In this case, the design of C-2 and probes tethered at other groups may be useful for the identification of other binding proteins.

The benzophenone (BZDC) group has largely replaced the arylazide (ASA) as the photoprobe of choice for many biochemical studies (150). It is chemically more stable, it does not decompose in ambient light, and it selectively modifies hydrophobic binding regions in target macromolecules (149). For example, $[^3H]BZDC-Ins(1,4,5)P_3$ was used to map the binding site in the rat brain $Ins(1,4,5)P_3$ receptor, when insufficient labelled protein was obtained with an $[^{125}I]$ASA-Ins$(1,4,5)P_3$ photoprobe (151). An entire family of photoaffinity probes (149) has been developed based on the efficacy of the BZDC-InsPn probes for selective covalent modification of target proteins in signalling pathways. The BZDC-Ins$(1,3,4,5)P_4$ and -InsP_6 probes have been used successfully to label several InsPn binding proteins, including synaptotagmin I, recombinant C2B domains, native Golgi COPs, native AP-2, recombinant AP-3, and a number of PtdIns$(3,4,5)P_3$ target proteins (Prestwich, unpublished results). Using a BZDC-Ins$(1,2,6)P_3$ label, a 55 kDa umbilical cord epithelial cell target was identified (A. Chaudhary and G.D. Prestwich, unpublished data).

A convenient method for determining the InsPn and PtdInsPn selectivity of binding domains for individual inositol polyphosphates or inositol lipids involves photoaffinity labelling a glutathione S-transferase (GST) fusion construct of the domain. We have labelled C2A and C2B domains of synaptotagmin II and II (152), and constructs derived from the neuronal assembly protein AP-3 (J.M. Rabinovich, unpublished results). For example, the GST-C2B domain was expressed in *E. coli* and purified on GST affinity beads. The eluted proteins is generally > 80% homogeneous and can be photoaffinity labelled with 0.05 to 0.50 μCi $[^3H]$photolabel per 5–10 mg protein. For labelling the C2B domain, we have employed $[^3H]$BZDC, tethered at P-1 and P-2, to Ins$(1,4,5)P_3$, Ins$(1,3,4,5)P_4$ and InsP_6, as well as $[^3H]$BZDC-PtdIns$(4,5)P_2$ and -PtdIns$(3,4,5)P_3$. The uses and advantages of the benzophenone photophore (149, 150), specifically the BZDC photolabel have been described (153).

Protocol 7.   Photoaffinity labelling

Equipment and reagents

- 12-, 48- or 96-well flat-bottom plastic tissue culture plates
- UV lamp (360 nm 30 W high-intensity, Cole Parmer)
- Autoradiographic film for detection of $[^3H]$, such as $[^3H]$Hyperfilm (Amersham)
- Fraction containing binding proteins (membranes, crude soluble, detergent solubilized membranes, partially purified, recombinant, or purified protein)
- Radiolabelled photoaffinity probe, e.g., $[^3H]$BZDC-Ins$(1,3,4,5)P_4$ (34 Ci mmol^{-1})
- Unlabelled ligand (e.g. Ins$(1,3,4,5)P_4$ prepared at 10 × concentration)

- Photolabelling buffer (PB): 25 mM Tris-HCl pH 7.4, 1 mM EDTA, and 1.0 mM K_2HPO_4
- 5X SDS sample buffer (7.5 ml 20% SDS, 3.75 ml 2-mercaptoethanol, 3.75 ml glycerol, 582 mg Trizma preset 7.0 (Sigma) in a final volume of 15 ml)
- SDS-PAGE gel: 10% or 7.5% running gel with 4% stacking gel
- SDS-PAGE Gel Fixer (methanol:H2O:glacial acetic acid 5:5:1)
- Entensify two-component fluorography system (Dupont/NEN)
- Gel dryer
- Autoradiography cassettes and pens

Method

1. In each well of the multi-well plate (on ice) place 5–30 μl of the protein sample in a final volume between 50–100 μl of PB.

2. Add 20–70 nM [^{3}H]BZDC-Ins(1,3,4,5)P_4 (final concentration) for the totals and for the non-specific samples, 1–10 μM of unlabelled ligand.

3. Incubate for 10 min on ice in the dark.

4. Expose to UV light for 1 h, keeping the plate on ice. Support the lamp between 1–10 cm from the sample.

5. Add SDS sample buffer to stop the reaction. The plate can be stored at −20 °C until ready to load samples onto SDS-polyacrylamide gel.

6. Load the samples into the gel lanes and electrophorese. In addition, load prestained molecular weight markers in a separate lane.

7. Fix proteins in the gel with SDS-PAGE gel fixer.

8. Fluorograph the gel using Entensify according to the manufacturer's protocol or use an alternative fluorography system and dry under vacuum in a gel dryer (1–2 h at 65 °C).

9. Mark the film with an autoradiography pen to help align the film on the gel after exposure. Expose the gel to film for 4–21 days.

10. Develop the film according to the manufacturer's instructions. Films can then be digitized to quantify the effects of displacement by Ins$P$ns and inhibitors.

11. As with the other binding assays, the non-specific binding is determined by including 1–10 μM unlabelled ligand in a separate sample. In this instance, the amount of labelling of the protein in the gel can be determined by densitometry and the value for non-specific labelling subtracted from the total.

Notes:
1. We do not routinely stain our gels, but run prestained markers instead. Other groups have stained gels with Coomassie blue before fluorography without substantial interference from the stain.
2. The optimal time course of irradiation (10–120 min at 4 °C) should be established for a given 360 nm UV light set-up.
3. Several parameters can be optimized and competitors can be evaluated. In a single experiment, the labelling of 5–20 samples (for 1–2 mini or regular sized gels) is routinely performed.

7. Identifying PtdIns$P$n binding proteins with Ins$P$n ligands

In the past, PtdIns$P$n binding has often been reported for purified proteins, but rarely for crude fractionated proteins. Possible reasons for this include (a) the difficulty in separation of the protein bound lipid from the free lipid and (b) the limited availability of radiolabelled and unlabelled lipid ligands. A widely used technique for separation of bound and free lipid utilizes size exclusion/gel filtration chromatography, similar to that described above for the spun-column separation. Alternatively, various spectroscopic techniques have been employed to assess effects of lipid binding to proteins. While these techniques are effective in mapping and characterizing binding sites in individual purified protein preparations, they are neither convenient for screening large numbers of samples, nor are they suitable for detecting binding in crude or partially purified fractions. The identification of novel binding proteins for PtdIns*P*ns from crude soluble and crude solubilized membrane fractions may be more easily achieved using the inositol head groups as probes. Centaurin α, phospholipase C$_\delta$ and synaptotagmin were purified this way. Thus, many of the techniques described here may also be applicable for identification of novel PtdIns*P*n binding proteins. Furthermore, new photoprobes, containing appropriate long-chain fatty acid groups (which correspond to the endogenous PtdIns*P*ns) are being designed and synthesized by a number of groups (see refs. 112 and 148, and see the Appendix at the end of this volume), and we anticipate that these probes will facilitate the photolabelling and affinity purification of novel PtdIns*P*n binding proteins in the future.

Acknowledgements

The authors would like to thank Drs. Craig Garner, Vytas Bankaitis, and Peter Cullen for helpful discussions during the preparation of this chapter. We have also benefited from many valuable discussions with colleagues and from unpublished information and preprints. This work was supported by NIMH grant (R29MH50102) and DDRC (P50HD32901) to ABT. Work at Stony Brook was supported by NIH grant (NS29632) to GDP. TRJ is supported by the Medical Research Council. LPHO was supported by a Comprehensive Minority Faculty Development Program Fellowship from UAB and an NIMH Research Supplement for Underrepresented Minorities.

References

1. Stephens, L., Radenberg, T., Thiel, U., Vogel, G., Khoo, K.H., Dell, A., Jackson, T.R., Hawkins, P.T., and Mayr, G.W. (1993). *J. Biol. Chem.*, **268**, 4009–1015.
2. Shears, S.B., Ali, N., Craxton, A., and Bembenek, M.E. (1995). *J. Biol. Chem.*, **270**, 10 489–10 497.
3. Martin, J.B., Bakker-Grunwald, T., and Klein, G. (1993). *Eur. J. Biochem.*, **214**, 711–718.
4. Laussmann, T., Eujen, R., Weisshuhn, C.M., Thiel, U., and Vogel, G. (1996). *Biochem. J.*, **315**, 715–720.
5. Berridge, M.J. (1993). *Nature*, **361**, 315–325.
6. Ferris, C.D. and Snyder, S.H. (1992). *Annu. Rev. Physiol.*, **54**, 469–488.
7. Kuno, M. and Gardner, P. (1987). *Nature*, **326**, 301–304.
8. Khan, A.A., Steiner, J.P., Klein, M.G., Schneider, M.F., and Snyder, S.H. (1992). *Science*, **257**, 815–818.

9. Mayrleitner, M., Schafer, R., and Fleischer, S. (1995). *Cell Calcium*, **17**, 141–153.
10. De Waard, M., Seager, M., Feltz, A., and Couraud, F. (1992). *Neuron*, **9**, 497–503.
11. Zwiller, J., Ogasawara, E.M., Nakamoto, S.S., and Boynton, A.L. (1988). *Biochem. Biophys. Res. Commun.*, **155**, 767–772.
12. Kanematsu, T., Takeya, H., Wantanabe, Y., Ozaki, S., Yoshida, M., Kaga, T., Iwanaga, S., and Hirata, M. (1992). *J. Biol. Chem.*, **267**, 6518–6525.
13. Davis, F.B., Davis, P.J., Lawrence, W.D., and Blas, S.D. (1991). *FASEB J.*, **5**, 2992–2995.
14. Fraser, C.L. and Sarnacki, P. (1991). *Am. J. Physiol.*, **31**, F411–F416.
15. Restrepo, D., Teeter, J.M., Honda, E., Boyle, A.G., Maracek, J.F., Prestwich, G.D., and Kalinoski, L. (1992). *Am. J. Physiol.*, **263**, C667–C673.
16. Siren, M., Linne, L., and Persson, L. (1991). In *Pharmacological effect of D-myo-inositol-1,2,6-trisphosphate (PP56)* (ed. A. Reitz). Vol. 463, pp. 103–114. ACS Symposium Series, Washington, D.C.
17. Sun, X.Y., Edvinsson, L., Wahlestedt, C., Yoo, H., and Hedner, T. (1995). *J. Cardiovasc. Pharm.*, **25**, 696 - 704.
18. Heilig, M., Edvinsson, L., and Wahlestedt, C. (1991). *Eur. J. Pharm.*, **209**, 27–32.
19. Communi, D., Vanweyenberg, V., and Erneux, C. (1995). *Cell. Signalling*, **7**, 643–650.
20. Irvine, R.F. (1992). In *Inositol phosphates and calcium signalling* (ed. J.W. Putney). Vol. 26, pp. 161–185. Raven Press, New York.
21. Malviya, A.N. (1994). *Cell Calcium*, **16**, 301–313.
22. Luckhoff, A. and Clapham, D.E. (1992). *Nature*, **355**.
23. Smith, P.M. (1992). *Biochem. J.*, **283**, 27–30.
24. Fadool, D.A. and Ache, B.W. (1994). *Proc. Natl. Acad. Sci. USA*, **91**, 9471–9475.
25. Shirakawa, H. and Miyazaki, S. (1995). *Cell Calcium*, **17**, 1–13.
26. Ivorra, I., Gigg, R., Irvine, R.F., and Parker, I. (1991). *Biochem. J.*, **273**, 317–321.
27. Vajanaphanich, M., Schultz, C., Rudolf, M.T., Wasserman, M., Enyedi, P., Craxton, A., Shears, S.B., Tsien, R.Y., Barrett, K.E., and Traynor-Kaplan, A. (1994). *Nature*, **371**, 14092–14097.
28. Xie, W., Kaetzel, M., Bruzik, K., Dedman, J., Shears, S., and Nelson, D. (1996). *J. Biol. Chem.*, **271**, 14092–14097.
29. Menniti, F.S., Oliver, K.G., Putney, J.W., and Shears, S.B. (1993). *Trends Biochem. Sci.*, **18**, 53–56.
30. Sasakawa, N., Sharif, M., and Hanley, M. (1995). *Biochem. Pharm.*, **50**, 137–146.
31. Sasakawa, N., Ferguson, J.E., Sharif, M., and Hanley, M.R. (1994). *Molec. Pharm.*, **46**, 380–385.
32. Shamsuddin, A. . and Yang, G.-Y. (1995). *Carcinogenesis*, **16**, 1975–1979.
33. Han, O., Failla, M.L., Hill, A.D., Morris, E.R., and Smith, J.C. (1995). *Proc. Soc. Exp. Biol. Med.*, **210**, 50–56.
34. Kitchen, E., Condliffe, A.M., Rossi, A.G., Haslett, C., and Chilvers, E.R. (1996). *Br. J. Pharm.*, **117**, 979–985.
35. Morrison, R.S., Shi, E., Kan, M., Yamaguchi, F., McKeehan, W., Rudnicka-Nawrot, M., and Palczewski, K. (1994). *In Vitro Cellular and Dev. Biol.*, **30A**, 783–789.
36. Cecconi, O., Nelson, R.M., Roberts, W.G., Hanasaki, K., Mannori, G., Schultz, C., Ulich, T.R., Aruffo, A., and Bevilacqua, M.P. (1994). *J. Biol. Chem.*, **269**, 15060–15066.

37. Kar, S., Quirion, R., and Parent, A. (1994). *Neuroreport*, **5**, 625–628.
38. Llinas, R., Sugimori, M., Lang, E.J., Morita, M., Fukuda, M., Miinobe, M., and Mikoshiba, K. (1994). *Proc. Natl. Acad. Sci. USA*, **91**, 12990–12993.
39. Rao, P.S., Lui, X., Das, D.K., Weinstein, G.S., and Tyras, D.H. (1991). *Ann. Thorac. Surg.*, **52**, 908–912.
40. Hawkins, P.T., Poyner, D.R., Jackson, T.R., Letcher, A.J., Lander, D.A., and Irvine, R.F. (1993). *Biochem. J.*, **294**, 934–934.
41. Voglmaier, S., Bembenek, M.E., Kaplin, A.I., Dorman, G., Olszewski, J.D., Prestwich, G.D., and Snyder, S.H. (1996). *Proc. Natl. Acad. Sci. USA*, **93**, 4305–4310.
42. Liscovitch, M. and Cantley, L.C. (1994). *Cell*, **77**, 329–334.
43. Janmey, P.A. (1994). *Annu. Rev. Phys.*, **56**, 169–191.
44. Liscovitch, M. and Cantley, L.S. (1995). Cell, **81**, 659–662.
45. De Camilli, P., Emr, S.D., McPherson, P.S., and Novick, P. (1996). *Science*, **271**, 1533–1539.
46. Brown, H.A., Gutowski, S., Moomaw, C.R., Slaughter, C., and Sternweis, P.C. (1993). *Cell*, **75**, 1137–1144.
47. Hammond, S.M., Altshuller, Y.M., Sung, T.C., Rudge, S.A., Rose, K., Engebrecht, J., Morris, A.J., and Frohman, M. (1995). *J. Biol. Chem.*, **270**, 29640–29643.
48. Liscovitch, M., Chalifa, V., Pertile, P., Chen, C.S., and Cantley, L.C. (1994). *J. Biol. Chem.*, **269**, 21403–21406.
49. Shaw, G. (1996). *Bioessays*, **18**, 35–46.
50. Fukami, K., Sawada, N., Takeshi, E., and Takenawa, T. (1996). *J. Biol. Chem.*, **271**, 2646–2650.
51. Rebecchi, M., Peterson, A., and McLaughlin, S. (1992). *Biochemistry*, **31**, 12742–12747.
52. Fantl, W.J., Escobedo, J.A., Martin, G.A., Turck, C.W., del Rosario, M., McCormick, F., and Willians, L.T. (1992). *Cell*, **69**, 413–423.
53. Wennstrom, S., Hawkins, P., Cooke, F., Hara, K., Yonezawa, K., Kasuga, M., Jackson, T., Claesson-Welsh, L., and Stephens, L. (1994). *Curr. Biol.*, **4**, 385–393.
54. Wymann, M. and Acaro, A. (1994). *Biochem. J.*, **298**, 517–520.
55. Kotani, K., Yonezawa, K., Hara, K., Ueda, H., Kitamura, Y., Sakaue, H., Ando, A., Chavanieu, A., Calas, B., Grigorescu, F., Nishiyama, M., Waterfield, M.D., and Kasuga, M. (1994). *EMBO J.*, **13**, 2213–2321.
56. Okada, T., Kawano, Y., Sakakibara, T., Hazeki, O., and Ui, M. (1994). *J. Biol. Chem.*, **269**, 3568–3573.
57. Hara, K., Yonezawa, K., Sakaue, H., Ando, A., Kotani, K., Kitamura, T., Kitamura, Y., Ueda, H., Stephens, L., Jackson, T.R., Hawkins, P.T., Dhand, R., Clark, A.E., Holman, G.D., Waterfield, M.D., and Kasuga, M. (1994). *Proc. Natl. Acad. Sci. USA*, **91**, 7415–7419.
58. Cheatham, B., Vlahos, C.J., Cheatham, L., Wang, L., Blenis, J., and Kahn, C.R. (1994). *Mol. Cell Biol.*, **14**, 4902–1911.
59. Joly, M., Kazlauskas, A., and Corvera, S. (1995). *J. Biol. Chem.*, **270**, 13225–13230.
60. Chung, J., Grammer, T.C., Lemon, K.P., Kazlauskas, A., and Blennis, J. (1994). *Nature*, **370**, 71–75.
61. Burgering, B.M.T. and Coffer, P.J. (1995). *Nature*, **376**, 599–602.

62. Franke, T.F., Yang, S., Chan, T.O., Datta, K., Kazlausdas, A., Morrison, D.K., Kaplan, D.R., and Tsichlis, P.N. (1995). *Cell*, **81**, 727–736.
63. Kohn, A.D., Kovacina, K.S., and Roth, R. (1995). *EMBO J.*, **14**, 4288–4295.
64. Kimura, K., Hattori, S., Kabuyama, Y., Shizawa, Y., Takayanagi, J., Nakamura, S., Toki, S., Matsuda, Y., Onodera, K., and Fukui, Y. (1994). *J. Biol. Chem.*, **269**, 18 961–18 967.
65. Jackson, T.R., Blader, I.J., Hammonds-Odie, L.P., Burga, C.R., Cooke, F., Hawkins, P.T., Wolf, A.G., Heldman, K.A., and Theibert, A.B. (1996). *J. Cell Science*, **109**, 289–300.
66. Lee, M.-H. and Bell, R.M. (1991). *Biochemistry*, **30**, 1041–1049.
67. Chauhan, A., Brockerhoff, H., Wisniewski, H.M., and Cauhan, V.P.S. (1991). *Arch. Biochem. Biophys.*, **287**, 283–287.
68. Huang, F.L. and Huang, K.-P. (1991). *J. Biol. Chem*, **266**, 8727–8733.
69. Nakanishi, H., Brewer, K.A., and Exton, J.H. (1993). *J. Biol. Chem.*, **268**, 13–18.
70. Singh, S.S., Chauhan, A., Brockerhoff, H., and Chauhan, V.P.S. (1993). *Biochem. Biophys. Res. Commun.*, **195**, 104–112.
71. Toker, A., Meyer, M., Reddy, K.K., Falck, J.R., Aneja, R., Aneja, S., Parra, A., Burns, D.J., Ballas, L.M., and Cantley, L.C. (1994). *J. Biol. Chem.*, **269**, 32358–32367.
72. Rameh, L.E., Chen, C.-S., and Cantley, L.C. (1995). *Cell*, **83**, 821–830.
73. Zhang, J., Falck, J.R., Reddy, K.K., Abrams, C.S., Zhao, W., and Rittenhouse, S.E. (1995). *J. Biol. Chem.*, **270**, 22 807–22 810.
74. Palmer, R.H., Dekker, L.V., Woscholski, R., Le Good, J.A., Gigg, R., and Parker, P.J. (1995). *J. Biol. Chem.*, **270**, 22 412–22 416.
75. Theibert, A.B., Supattapone, S., Worley, P.F., Baraban, J.M., Meek, J.L., and Snyder, S.H. (1987). *Biochem. Biophys. Res. Commun.*, **148**, 1283–1289.
76. Bradford, P.G. and Irvine, R.F. (1987). *Biochem. Biophys. Res. Commun.*, **149**, 680–685.
77. Enyedi, P. and Williams, G.H. (1988). *J. Biol. Chem.*, **263**, 7940–7942.
78. Enyedi, P., Brown, E., and Williams, G.H. (1989). *Biochem. Biophys. Res. Commun.*, **159**, 200–208.
79. Donie, F. and Reiser, G. (1989). *FEBS Lett.*, **254**, 155–158.
80. Challis, R.A., Willcocks, A.L., Mulloy, B., Potter, B.V., and Nahorski, S.R. (1991). *Biochem. J.*, **274**, 861–867.
81. Cullen, P.J. and Irvine, R.F. (1992). *Biochem. J.*, **288**, 149–154.
82. Zhang, L., Bradley, M.E., Khoyi, M., Westfall, D.P., and Buxton, I.L. (1993). *Br. J. Pharm.*, **109**, 905–912.
83. Cullen, P.J., Patel, Y., Kakkar, V.V., Irvine, R.F., and Authi, K.S. (1994). *Biochem. J.*, 739–742.
84. Parent, A. and Quirion, R. (1994). *Eur. J. Neurosci.*, **6**, 67–74.
85. Hawkins, P.T., Reynolds, D.J.M., Poyner, D.R., and Hanley, M.R. (1990). *Biochem. Biophys. Res. Commun.*, **167**, 819–827.
86. Nicoletti, F., Bruno, V., Cavallaro, S., Copani, A., Sortino, M.A., and Canonico, P.L. (1990). *Mol. Pharmacol.*, **37**, 689–693.
87. Day, N.S., Ghalayini, A.J., and Anderson, R.E. (1995). *Curr. Eye Res.*, **14**, 851–855.
88. Yoo, H., Fallgren, B., Lindahl, A., and Wahlestedt, C. (1994). *Eur. J. Pharm.*, **268**, 55–63.
89. Stricker, R., Westerberg, E., and Reiser, G. (1996). *Br. J. Pharm.*, **117**, 919–925.

90. Walsh, D.A., Mapp, P.I., Polak, J.M., and Blake, D.R. (1995). *J. Pharm. Exp. Thera.*, **273**, 461–469.

91. Isaacks, R.E. and Harkness, D.R. (1980). *Am. Zool.*, **20**, 115–129.

92. Thieleczek, R., Mayr, G.W., and Brandt, N.R. (1989). *J. Biol. Chem.*, **264** (13), 7349–7356.

93. Spychala, J. (1987). *Biochem. Biophys. Res. Commun.*, **148**, 106–111.

94. Palczewski, K., Pulvermuller, A., Buczylko, J., Gutmann, C., and Hofmann, K.P. (1991). *FEBS Lett.*, **295**, 195–199.

95. Supattapone, S., Worley, P.F., Baraban, J.M., and Snyder, S.H. (1988). *J. Biol. Chem.*, **263**, 1530–1534.

96. Keen, J. (1990). *Annu. Rev. Biochem.*, **59**, 415–438.

97. Voglmaier, S.M., Keen, J.H., Murphy, J.-E., Ferris, C.D., Prestwich, G.D., Snyder, S.H., and Theibert, A.B. (1992). *Biochem. and Biophys. Res. Comm.*, **187**, 158–163.

98. Timerman, A.P., Mayrleitner, M.M., Lukas, T.J., Chadwick, C.C., Saito, A., Watterson, D.M., Schindler, H., and Fleischer, S. (1992). *Proc. Natl. Acad. Sci. USA*, **89**, 12655–12662.

99. Beck, K.A. and Keen, J.H. (1991). *J. Biol. Chem.*, **266**, 4442–4447.

100. Gaidarov, I.O. and Keen, J.H. (1995). *Mol. Biol. of Cell*, **6**S, 407.

101. Hao, W., Tan, Z., Prasad, K., Reddy, K.K., Chen, J., Prestwich, G.D., Falck, J.R., Shears, S.B., and Lafer, E.M. (1997). *J. Biol. Chem.*, in press.

102. Norris, F.A., Ungewickell, E., and Majerus, P.W. (1995). *J. Biol. Chem.*, **270**, 214–217.

103. Ye, W., Ali, N., Bembenek, M., Shears, S.B., and Lafer, E. (1995). *J. Biol. Chem.*, **270**, 1564–1568.

104. Morris, S.A., Ahle, S., and Ungewickell, E. (1989). *Curr. Opin. Cell Biol.*, **1**, 684–690.

105. Trowbridge, I. (1991). *Curr. Opin. Cell Biol.*, **3**, 634–641.

106. Robinson, M. (1987). *J. Cell Biol.*, **104**, 887–895.

107. Malhotra, V., Serafini, T., Prco, L., Shepherd, J.C., and Rothman, J.E. (1989). *Cell*, **58**, 329–336.

108. Rothman, J.E. (1994). *Nature*, 372, 55–63.

109. Rothman, J.E. and Wieland, F.T. (1996). *Science*, **272**, 227–229.

110. Fleischer, B., Xie, J., Mayrleitner, M., Shears, S.B., and Palmer, D.J. (1994). *J. Biol. Chem.*, **269**, 17826–17832.

111. Ali, N., Duden, R., Bembenek, M.E., and Shears, S.B. (1995). *Biochem. J.*, **310**, 279–284.

112. Prestwich, G.D. (1996). *Accounts of Chemical Research*, **29**, 503–513.

113. Matthew, W.D., Tsavaler, L., and Reichardt, L.F. (1981). *Jour. Cell Biol.*, **91**, 257–269.

114. Sudhof, T.C. and Jahn, R. (1991). *Neuron*, **6**, 665–677.

115. Niinobe, M., Yamaguchi, Y., Fukuda, M., and Mikoshiba, K. (1994). *Biochem. Biophys. Res. Commun.*, **205**, 1036–1042.

116. Fukuda, M.J., Aruga, M., Niinobe, S., Aimoto, S., and Mikoshiba, K. (1994). *J. Biol. Chem.*, **269**, 29206–29211.

117. Fukuda, M., Moreira, J.E., Lewis, F.M.T., Sugimori, M., Niinobe, M., Mikoshiba, K., and Llinas, R. (1995). *Proc. Natl. Acad. Sci. USA.*, **92**, 10708–10712.

118. Perin, M.S., Fried, V.A., Mignery, G.A., Jahn, R., and Sudhof, T.C. (1990). *Nature*, **345**, 260–263.
119. Bommert, K., Charlton, M.P., DeBello, W.M., Chin, G.J., Betz, H., and Augustine, G.J. (1993). *Nature*, **363**, 163–165.
120. Donie, F. and Reiser, G. (1991). *Biochem. J.*, **275**, 453–457.
121. Reiser, G., Kunzelmann, R., Hulser, E., Stricker, R., Hoppe, J., Lottspeich, F., and Kalbacccher, H. (1994). *Biochem. Biophys. Res. Commun.*, **214**, 20–27.
122. Stricker, R., Kalbacher, H., Lottspeich, F., and Reiser, G. (1995). *FEBS Lett.*, **370**, 236–240.
123. Cullen, P.J., Chung, S.K., Chang, Y.T., Dawson, A.P., and Irvine, R.F. (1995). *FEBS Lett.*, **358**, 240–242.
124. Cullen, P.J., Dawson, A.P., and Irvine, R.F. (1995). *Biochem. J.*, **305**, 139–143.
125. Cullen, P.J., Hsuan, J.J., Troung, O., Letcher, A.J., Jackson, T.R., Dawson, A.P., and Irvine, R.F. (1995). *Nature*, **376**, 527–530.
126. Cunningham, E.B. (1995). *Biochem. Biophys. Res. Commun.*, **215** (1), 212–218.
127. Koppler, P., Mersel, M., Humbert, J.-P., Vignon, J., Vincendon, G., and Malviya, A.N. (1996). *Biochemistry*, **35**, 5481–5487.
128. Kanematsu, T., Misumi, Y., Watanabe, Y., Ozaki, S., Koga, T., Iwanaga, S., Ikehara, Y., and Hirata, M. (1996). *Biochem. J.*, **313**, 319–325.
129. Hammonds-Odie, l.P., Jackson, T.R., Profit, A.A., Blader, I.J., Turck, C.W., Prestwich, G.D., and Theibert, A.B. (1996). *J. Biol. Chem.*, **271**, 18859–18868.
130. Theibert, A.B., Estevez, V.A., Ferris, C.D., Danoff, S.K., Barrow, R.K., Prestwich, G.D., and Snyder, S.H. (1991). *Proc. Natl. Acad. Sci. USA*, **88**, 3165–3169.
131. Theibert, A.B., Estevez, V.A., Mourey, R.J., Marecek, J.F., Barrow, R.K., Prestwich, G.D., and Snyder, S.H. (1992). *J. Biol Chem.*, **267**, 9071–9079.
132. Reiser, G., Kunzelmann, U., Hulser, E., Stricker, R., Hoppe, J., Lottspeich, F., and Kalbacher, H. (1995). *Biochem. Biophys. Res. Commun.*, **214**, 20–27.
133. Nishizuka, Y. (1988). *Nature*, **334**, 661–665.
134. Irvine, R.F. (1996). *Nature*, **380**, 581–583.
135. Bourne, H.R., Sanders, D.A., and McCormick, R. (1991). *Nature*, **349**, 117–126.
136. Coleman, D.E. and Sprang, S.R. (1996). *Trends Biochem. Sci.*, **21**, 41–44.
137. Schiavo, G., Gmachl, M.J., Stenbeck, G., Sollner, T.H., and Rothman, J.E. (1995). *Nature*, **378**, 733–736.
138. Luttrell, B.M. (1994). *Cellular Signalling*, **6** (4), 355–362.
139. Ali, N. and Agrawal, D.K. (1992). *J. Pharm. Tox. Methods*, **27**, 79–83.
140. Kimura, Y., Kanematsu, T., Watanabe, Y., Ozaki, S., Koga, T., and Hirata, M. (1991). *Biochim. Biophys. Acta.*, **1069**, 218–222.
141. Malviya, A.N., Rogue, P., and Vincedon, G. (1990). *Proc. Natl. Acad. Sci. USA*, **87**, 9270–9274.
142. Crawford, N., Hack, N., and Authi, K.S. (1992). In *Methods in Enzymology* (ed. J.J. Hawiger) Vol. 245, pp. 5–20. Academic Press, Inc., San Diego.
143. Gordon-Weeks, P.R. (1987). In *Neurochemistry: a practical approach.* (ed. A.J. Turner and H.S. Bachelard), pp. 1–26. IRL Press, Oxford.
144. Worley, P.F., Baraban, J.M., Colvin, J.S., and Snyder, S.H. (1987). *Nature*, **325**, 159–161.
145. Chadwick, C.C., Saito, A., and Fleischer, S. (1991). *Biophys. J.*, **59**, 525a-.
146. Cullen, P.J., Comerford, J.G., and Dawson, A.P. (1988). *FEBS Lett.*, **228**, 57–59.

147. Prestwich, G.D., Marecek, J.F., Mourey, R.J., Theibert, A.B., Ferris, C.D., Danoff, S.K., and Snyder, S.H. (1991). *J. Am. Chem. Soc.*, **113**, 1822–1825.
148. Reiser, G., Schafer, R., Donie, F., Hulser, E., Nehls-Nafabandu, M., and Mayer, G.W. (1991). *Biochem. J.*, **280**, 533–539.
149. Prestwich, G.D., Dorman, G., Elliot, J.T., Marecak, D.M., and Chaudhary, A. (1997). *Photochem. and Photobiol.*, in press.
150. Dorman, G. and Prestwich, G.D. (1994). *Biochemistry*, **33**, 5661–5673.
151. Mourey, R.J., Estevez, V.A., Marecek, J.F., Barrow, R.K., Prestwich, G.D., and Snyder, S.H. (1993). *Biochemistry*, **32**, 1719–1726.
152. Mehrotra, B., Elliott, J.T., Chen, J., Olszewski, J.D., Profit, A.A., Chaudhary, A., Fukuda, M., Mikoshiba, K., and Prestwich, G.D. (1997). *J. Biol. Chem.*, in press.
153. Olszewski, J.D., Dorman, G., Elliot, J.T., Hong, Y., Ahern, D.G., and Prestwich, G.D. (1995). *Bioconj. Chem.*, **6**, 395–400.
154. Schiavo, G., Gu, Q.-M., Prestwich, G.D., Söllner, T.H., and Rothman, J.E. (1996) *Proc. Natl. Acad. Sci. USA*, **93**, 13327–13332.

Mass assay of inositol 1,4,5-trisphosphate and phosphatidylinositol 4,5-bisphosphate

R. A. JOHN CHALLISS

1. Introduction

The phosphoinositides are now established as a major source of second messenger molecules involved in intracellular signalling. Furthermore, as well as fulfilling this precursor function, the (poly)phosphoinositides are increasingly recognized as being important signalling molecules in their own right. Thus, the hydrolysis of phosphatidylinositol 4,5-bisphosphate (PtdIns(4,5)P_2) by phospholipase C to generate the second messengers inositol 1,4,5-trisphosphate (Ins(1,4,5)P_3) and sn-1,2-diacylglycerol (DAG), or the phosphorylation of PtdIns(4,5)P_2 by PtdIns(4,5)P_2 3-kinase to generate phosphatidylinositol 3,4,5-trisphosphate (PtdInsP_3), can both be activated by an array of hormones, neurotransmitters or other signalling molecules. In addition, the change in PtdIns(4,5)P_2 concentration which may accompany PLC activation can have a regulatory effect on other important enzymes whose activity or cellular localization is influenced by PtdIns(4,5)P_2 (1).

Elucidation of the phosphoinositide cycle (by which PtdIns(4,5)P_2 is hydrolysed, and Ins(1,4,5)P_3 and DAG are metabolized for re-incorporation into cellular the phosphoinositide pool), and studies which have highlighted the emerging roles of PtdIns(3,4,5)P_3 and higher inositol polyphosphates in cellular regulation, have been achieved largely by employing radioisotopic labelling strategies. Although radiolabelling methods continue to be invaluable for addressing many of the unresolved questions concerning phosphoinositides and cellular regulation, methods for the mass determination of key cycle intermediates can offer a number of clear advantages in terms of speed/throughput, reproducibility and cost. In this chapter methods for the mass assay of two critical phosphoinositide cycle intermediates are described.

2. Preparation of an Ins(1,4,5)P_3 binding protein

The Ins(1,4,5)P_3 receptor exhibits both high affinity and stereo- and positional specificity for Ins(1,4,5)P_3 and thus satisfies the principal criteria defining an experimentally useful Ins(1,4,5)P_3- recognition protein (2–4). Although the richest natural source of the Ins(1,4,5)P_3-receptor is cerebellum, the binding proteins prepared from either rat, pig or cow cerebellum exhibit higher K_D values for Ins(1,4,5)P_3 binding that are observed in other preparations (e.g. bovine adrenal cortex). In practical terms this means that more care is required to preserve the Ins(1,4,5)P_3-bound/cerebellar receptor complex, making some methods of separating bound and free Ins(1,4,5)P_3 (e.g. filtration) unreliable, although some workers have found that the use of a more alkaline pH (8.3) can decrease the K_D for Ins(1,4,5)P_3 binding sufficiently to allow filtration to be used to separate bound and free ligand (5). Notwithstanding this caveat, a crude 'P$_2$' fraction isolated from bovine adrenal cortex is recommended as the source of the Ins(1,4,5)P_3-receptor (see Protocol 1), although alternatives (in terms of species and tissue) are likely to exist. Indeed, any tissue obtainable in abundance and yielding a relatively high affinity receptor (K_D 5 nM) and sufficiently high Ins(1,4,5)P_3-receptor density (ideally B$_{max.}$ $\geqslant$ 1 pmol mg^{-1} membrane protein) in a crude 'P$_2$' fraction should suffice.

Protocol 1. Bovine adrenal cortical Ins(1,4,5)P_3-binding protein

Equipment and reagents

- Basic dissection equipment: scissors, scalpel, forceps, spatula
- Homogenizer: either a Polytron (Brinkmann Instruments, Westbury, NY) or and Ultra-Turrax (Jankel & Kunkel KG, Staufen i. Breisgau, Germany)
- Homogenization buffer: 20 mM NaHCO$_3$, 1 mM dithiothreitol, pH 8.0
- Refrigerated centrifuge capable of 50 000 × g and handling suitable volumes (at least 8 × 50 ml centrifuge tubes)

A. *Tissue dissection*

1. Place on ice within 15 min of slaughter intact adrenal glands obtained from an abattoir. Cleaned intact glands can be stored at −80 °C for at least 6 months without compromising the yield or quality of the Ins(1,4,5)P_3-binding protein subsequently prepared.

2. Cut the glands (fresh or thawed completely on ice from frozen) longitudinally to reveal the inner medulla and outer cortex structure. Remove the inner medulla using either scissors or a scalpel (care should be taken to remove > 90% of the medulla material) and scrape off the remaining cortex from the outer capsule using a spatula. Add the cortex tissue to a pre-weighed container in an ice-bath until all glands have been dissected.

(N.B. 8 bovine glands will yield ~60 g of adrenal cortex which will produce ~60 ml of the final binding protein preparation, sufficient for 2000 assay samples).

B. *Preparation of binding protein*

1. Dispense approximately 10 g portions of adrenal cortex into 50 ml centrifuge tubes, add about 20 ml homogenization buffer and thoroughly homogenize the cortex (e.g. 5 × 20 sec bursts of Polytron at setting 5–6) preferably whilst constantly maintained on ice. Bring all tubes to 50 ml by further addition of homogenization buffer, then balance and centrifuge at 4°C for 10 min at 5000 × g.

2. Decant the supernatant into a beaker in an ice-bath and re-extract the pellet by addition of further homogenization buffer. Then homogenize, centrifuge and recover the supernatant fraction as described above.

3. Decant the pooled 'S$_1$' fraction into clean centrifuge tubes and centrifuge at 4°C for 20 min at 40000 × g. At this stage discard the supernatant, re-homogenize the 'P$_2$' pellet in homogenization buffer (50 ml final volume) and re-centrifuge at 4°C for 20 min at 40000 × g. Repeat this latter stage twice more and take up the final washed 'P$_2$' pellet by homogenization in a small volume (5–8 ml) of homogenization buffer. Combine the 'P$_2$' fractions and perform a protein assay at this stage to allow the protein concentration to be adjusted to 15–18 mg ml^{-1}.

4. Dispense the final Ins(1,4,5)P_3-binding protein into suitable aliquots (e.g. a 3 ml aliquot provides sufficient binding protein for a standard curve plus 36 samples assayed in duplicate). Aliquots should be frozen and stored at −20°C or −80°C. Refreezing partially used aliquots should be avoided. The preparation is stable at −20°C for at least 6 months at least in terms of its Ins(1,4,5)P_3 recognition properties.

3. Sample preparation for Ins(1,4,5)P_3 mass assay

It is impossible to provide a single protocol for preparing samples for the mass assay as details will vary depending on whether the biological samples are intact tissues or dissociated/cultured cells, and whether the cells are grown on a substratum or are present in suspension. As a rule of thumb, established termination methods applicable for the extraction and measurement of other rapidly turning-over cellular metabolites (e.g. ATP, NAD, cyclic nucleotides, etc.) should be employed.

For termination/extraction of intact tissues, rapid freezing in liquid N$_2$ followed by tissue disruption (using either a pestle and mortar, or a Polytron/Ultra Turrax homogenizer) in the presence of either perchloric acid or trichloroacetic acid is recommended. For cells (and thin tissue slice preparations) addition of ice-cold perchloric or trichloroacetic acid to a final concentration of ~0.5 M is usually sufficient to arrest instantly metabolism and cause the release of Ins(1,4,5)P_3 into the acid-medium. However, there are cases where more radical disruption methodologies are necessary (e.g. sonication) to elicit total Ins(1,4,5)P_3 release (6). When establishing a protocol, the extent of inositol (poly)phosphate release should be checked using [^{3}H]inositol-labelled cells. A typical procedure for preparing samples for mass assay from confluent monolayers of cultured cells is given in *Protocol 2*.

Protocol 2. Sample preparation for Ins(1,4,5)P_3 mass assay

Reagents
- 1 M trichloroacetic acid
- water-saturated diethyl ether, or freshly prepared freon/tri-*n*-octylamine (1:1, v/v)
- 60 mM NaHCO$_3$; 30 mM EDTA (adjusted to pH 7 with NaOH)

A. *Termination*

1. Terminate the incubation by adding an equal volume of ice-cold 1 M trichloroacetic acid (TCA). Alternatively, carefully aspirate the incubation medium and add sufficient ice-cold 0.5 M TCA to completely cover the monolayer of cells (e.g. 250 μl per well of a standard 24-well multiwell plate).

2. As quickly as possible, transfer the plate to an ice-bath and allow extraction to occur for 15–20 min. In processing multiple samples on a single plate it is inevitable that acid-treated cell monolayers will remain in the incubator/water bath exposed to an ambient temperature of 37 °C. However, the time of this exposure should be minimized by ordering experiments in such a way that termination of each individual well occurs over as shorter time as possible.

B. *Neutralisation of TCA-stopped samples*

1. Carefully transfer each acid supernatant to a solvent-resistant disposal 5 ml tube. (This should leave monolayers on the multiwell plate devoid of the original aqueous-soluble cellular contents but unaffected with respect to membrane lipids. These can be recovered by subsequent addition of acidified chloroform/methanol; see *Protocol 4*).

2. Wash the acid-supernatant four times with three volumes of water-saturated diethyl ether. (This should be performed in a fume-hood.) Following each addition, thoroughly vortex-mix samples to achieve maximum transfer of TCA from the aqueous into the diethyl ether phase. Following the final wash, aspirate all of the diethyl ether phase and leave samples for 30–45 min in the fume-hood to achieve complete removal of diethyl ether.

3. Transfer a volume of the neutralized supernatant (200 μl) to an Eppendorf tube and add 30 mM EDTA (50 μl) and 60 mM NaHCO$_3$ (50 μl), and then vortex-mix and cap the samples. At this stage samples can be stored at 4°C for up to 14 days prior to mass assay.

C. *An alternative neutralisation method*

1. Carefully transfer acid supernatant samples to Eppendorf tubes, and add 1 volume of 10 mM EDTA (pH 7) per 4 volumes of sample. Add an equal volume of freshly prepared freon/tri-*n*- octylamine (1:1, v/v) (see ref. 7), and cap and thoroughly vortex-mix each tube (3 × 10 sec).

2. Centrifuge the samples (10 000 × *g*, 5 min) and recover a known volume of the uppermost phase into a new Eppendorf tube. Bring the pH to 7 by the addition of 1 volume of 60 mM NaHCO$_3$ per 4 volumes of sample. Samples can then be stored at 4°C for up to 14 days prior to mass assay. Although this method appears more convenient that the multiple diethyl ether wash procedure outlined above, the final neutral extract is often contaminated by traces of freon/tri-*n*-octylamine which can make subsequent pipetting of small aliquots less easy and more subject to inaccuracy.

D. *Generation of a 'buffer-blank'*

1. A 'buffer-blank' should be prepared as the diluent for the Ins(1,4,5)P_3 standard to allow construction of a suitable standard curve for the mass assay (see Protocol 3). This is achieved by making-up a large volume (~8 ml) of either acid or acidified medium and processing it according to the appropriate procedure described under B or C above.

4. Mass determination of Ins(1,4,5)P_3

The impressive stereo- and positional specificity exhibited by the Ins(1,4,5)P_3-receptor (8) means that other naturally occurring and closely structurally related inositol polyphosphates (e.g. Ins(1,3,4)P_3, Ins(1,3,4,5)P_4, Ins(1,4)P_2) bind to this receptor with much lower affinity and therefore only Ins(1,4,5)P_3 will be present in cell extracts at a sufficient concentration to displace [^{3}H]Ins(1,4,5)P_3 from the Ins(1,4,5)P_3-binding protein (4, 9). An Ins(1,4,5)P_3 mass assay based on the bovine adrenal cortical binding protein has been widely employed since it was first described (2, 10). This assay is conducted at an optimal pH (8.0) for maximal specific Ins(1,4,5)P_3 binding, at temperatures as close to 0°C as is practicable, and in the presence of an excess of EDTA. Under these conditions less than 1% of the added [^{3}H]Ins(1,4,5)P_3 is metabolized over a 2 h incubation period, despite the presence of a particulate Ins(1,4,5)P_3 5-phosphatase activity in the 'P$_2$' fraction

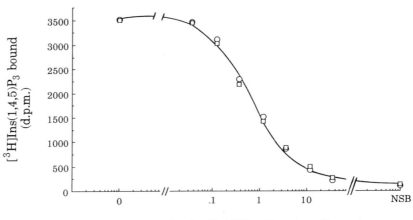

Figure 1 A typical standard curve for the Ins(1,4,5)P_3 mass assay. Duplicate samples of the indicated concentrations of Ins(1,4,5)P_3 (made up in an appropriate buffer-blank) were mixed with the assay buffer and [^{3}H]Ins(1,4,5)P_3 (8900 d.p.m. per assay; 20 Ci mmol^{-1}) at 4°C. Non-specific binding (NSB) was determined in the presence of 10 μM Ins(1,4,5)P_3. Incubations were started by addition of ~425 μg of the bovine adrenal cortex 'P$_2$' fraction bringing the final assay volume to 120 μl. After 30 min bound and free [^{3}H]Ins(1,4,5)P_3 were separated by rapid vacuum filtration and filter discs extracted 24 h before scintillation counting (see *Protocol 3*).

which can be revealed by increasing the temperature, omitting EDTA and adding 1 mM MgCl$_2$ to the assay mixture.

The K_D of Ins(1,4,5)P_3 for the adrenal cortical Ins(1,4,5)P_3-receptor is generally reported to be 2–5 nM (3), an affinity which allows separation of bound and free [^{3}H]Ins(1,4,5)P_3 using either rapid vacuum filtration or centrifugation. The former method is recommended as it yields highly reproducible results, both within and between assays, and has the advantage of reducing the non-specific binding component to < 5% of total binding (see *Figure 1*). However, to achieve such results a number of precautions must be observed. In particular, reagents, the assay mixture and crucially the wash buffer (see *Protocol 3*) must be maintained ice-cold at all times. Furthermore, the filtration procedure must be optimized as > 10% dissociation of equilibrium bound [^{3}H]Ins(1,4,5)P_3 can occur following addition of the wash buffer if filtration is delayed by as little as 20–30 sec (11, 12). In the light of these comments it should be emphasized that semi-automated procedures (e.g. using a Brandel cell harvester) need to be conducted in a cold-room (4°C) and special care must be taken to ensure that all solutions are maintained at 0–4°C.

The Ins(1,4,5)P_3 mass assay using vacuum filtration to separate bound and free ligand routinely used in my laboratory is described in *Protocol 3*, which also includes details of the use of centrifugation as a separation method. A variation on this method which allows the assay of larger volumes of cell extract has been reported (3). Note that non-specific binding values obtained using the filtration method are extremely low (< 5% of total binding). Unknown values can be obtained using a standard curve (see *Protocol 3A* and *Figure 1*), either by linearizing the curve (e.g. by plotting specific binding versus log [Ins(1,4,5)P_3]) and calculation using the best fit linear regression equation, or by using an iterative curve-fitting programme such as InPlot (GraphPad Software, San Diego, CA).

Protocol 3. The Ins(1,4,5)P_3 mass assay

Equipment and reagents

- For filtration: either Millipore 12-port vacuum manifolds (e.g. 1225 Sampling Manifold; Millipore Intertech, Bedford, MA) linked to a suitable vacuum pump; or a Brandel Cell Harvester (Brandel Biomedical Research & Development Laboratories, Gaithersburg, MA). Whatman GF/B or GF/C filter discs or sheets.

  For centrifugation: a refrigerated bench-top centrifuge capable of holding 12 samples and rotating at 10 000 r.p.m.
- Assay buffer: 100 mM Tris/HCl, 4 mM EDTA, pH 8.0
- [^{3}H]Ins(1,4,5)P_3 (DuPont NEN: NET-911, 15–30 Ci mmol^{-1}; or Amersham International: TRK 999, 20–60 Ci mmol^{-1}) diluted in distilled water to give approximately 222 000 d.p.m. ml^{-1} (i.e. 0.1 μCi ml^{-1})

- D-*myo*-Ins(1,4,5)P_3 (RBI Cat. No. I-105, Research Biochemicals Inc., Natick, MA; or CellSignals Inc., Lexington, KY) made up in distilled water and stored in 1 mM (20 μl) aliquots at −20 or −80°C. More dilute stock solutions (e.g. 40 μM) can be stored at −20°C for at least 3 months but should not be re-frozen
- Bovine adrenal cortex Ins(1,4,5)P_3-binding protein prepared as described in *Protocol 1*
- Wash buffer for filtration assays: 25 mM Tris/HCl, 1 mM EDTA, 5 mM NaHCO$_3$, pH 8.0; this solution should be placed in a freezer before use for a sufficient period for ice crystals to appear (i.e. it should be used at a temperature as close to 0°C as is practicable)

A. *Standards*

1. A suitable standard curve can be generated using final assay concentrations of Ins(1,4,5)P_3 of 0.3–300 nM; in a final assay volume of 120 μl these equate to 0.036–36 pmol Ins(1,4,5)P_3 per assay.

2. Dilute 20 μl aliquot of 1 mM Ins(1,4,5)P_3 with 480 μl of the 'buffer-blank' (see *Protocol 2*) to give a 40 μM solution. Duplicate or triplicate 30 μl aliquots of this are used to define non-specific binding (NSB; i.e. 10 μM final concentration in the assay volume of 120 μl). Serial dilutions of the 40 μM Ins(1,4,5)P_3 solution are performed to generate 1.2, 4, 12, 40, 120, 400, and 1200 nM solutions which generate 0.3 –300 nM final assay concentrations of Ins(1,4,5)P_3.

B. *Performing the assay*

1. All solutions should be maintained at 0–4°C throughout the assay. Aliquot 30 μl of either standard Ins(1,4,5)P_3 solution (0.036–36 pmol Ins(1,4,5)P_3 per assay in duplicate, plus 'buffer-blank' and NSB samples)

Protocol 3. *Continued*

or unknown sample generated as in *Protocol 2*, into disposable 3.5 or 5 ml test tubes (filtration) or Eppendorf tubes (centrifugation).

2. Add 30 μl of assay buffer (100 mM Tris/HCl, 4 mM EDTA, pH 8.0) and 30 μl of [^{3}H]Ins(1,4,5)P_3 (diluted as described above to give 6500–8000 d.p.m./assay).

3. The adrenal cortex Ins(1,4,5)P_3-binding protein should be thawed completely on ice. If the suspension has become 'clumpy' during the freezing/thawing process it should be homogenized immediately before use, although repeatedly titurating the solution with a pipette is usually sufficient to produce a homogenous preparation.

4. Add 30 μl of the adrenal cortex Ins(1,4,5)P_3-binding protein to each tube and vortex-mix thoroughly, ensuring that the entire 120 μl final volume is at the base of the tube.

5. Samples should be left on ice for at least 30 min to allow equilibrium Ins(1,4,5)P_3 binding to occur. Samples can be left for at least 2 h without significant changes in the [^{3}H]-Ins(1,4,5)P_3-binding values obtained.

6a. For separation of bound and free ligand by filtration:
    Load manifolds with filter discs and wet by the addition of 3–5 ml ice-cold wash-buffer (25 mM Tris/HCl, 1 mM EDTA, 5 mM NaHCO$_3$, pH 8.0). Then dilute an Ins(1,4,5)P_3 assay sample with 3 ml wash-buffer which is then immediately poured onto a filter. Wash the tube rapidly with 3 ml wash-buffer and wash the manifold well/filter disc with a further 3 ml of wash-buffer. The entire process (i.e. addition of 3 × 3ml wash-buffer for each sample) should be completed in 5–10 sec. When all samples have been filtered by this procedure transfer the filter discs to scintillation vials, and add sufficient scintillant to completely cover the disc. Irrespective of the scintillant used, samples should be left for a suitable time for the disintegrations detected by the scintillation counter to reach a constant maximal level (typically 12–24 h).

6b. For separation of bound and free ligand by centrifugation:
    Centrifuge samples at 10 000 × *g* for 3–5 min at 4 °C. It is preferable to use an angled-rotor which favours the formation of a tightly-packed pellet. Immediately transfer samples back to an ice-bath and remove the supernatant manually using a pipette, or gently aspirate them. The need to use a relatively large amount of the adrenal cortex Ins(1,4,5)P_3-binding protein (450–600 μg protein per assay) may result in a pellet which prevents the clean separation of the supernatant and therefore this method will yield more variable data. A number of techniques have been attempted in order to achieve a clearer separation

158

of bound and free [^{3}H]Ins(1,4,5)P_3, including the introduction of a sucrose 'cushion' which partitions between the pellet and supernatant phases (13), and increasing the salt concentration (e.g. by addition of NaCl and/or KCl) in the assay buffer (see ref. 12). Irrespective of the manner in which a complete separation of bound and free [^{3}H]Ins(1,4,5)P_3 is achieved, the pellet should be dissolved using 100 μl of 2% SDS at room temperature for 24 h. Following thorough vortex-mixing to completely solubilize the pellet, add 1 ml of scintillant and count the capped Eppendorf.

5. Sample preparation and processing for determination of PtdIns(4,5)P_2 mass

Established methods are available to isolate membrane phospholipids from cells and tissues, and to discriminate between the different phospholipid species by release and quantitation of the polar headgroup (14). Alkaline hydrolysis of inositol phospholipids will release Ins(1)P, Ins(1,4)P_2 and Ins(1,4,5)P_3 respectively from PtdIns, PtdIns(4)P and PtdIns(4,5)P_2. However, by-products will also be generated as the result of chemical hydrolysis. In the case of PtdIns(4,5)P_2, Ins(4,5)P_2 and Ins(2,4,5)P_3 will be generated as significant minor products in addition to Ins(1,4,5)P_3 (15). Of these three inositol polyphosphates only Ins(1,4,5)P_3 binds to the Ins(1,4,5)P_3-receptor with high affinity (9). Therefore, if the relative proportions of Ins(1,4,5)P_3, Ins(4,5)P_2 and Ins(2,4,5)P_3 generated under standard alkaline hydrolysis conditions can be determined (see *Protocol 4* and *Figure 2*), and remain essentially constant under conditions where the cellular level of PtdIns(4,5)P_2 may vary (e.g. agonist-challenge), the mass of PtdIns(4,5)P_2 in cells and tissues can be determined using the Ins(1,4,5)P_3 assay described above.

Total membrane phospholipid extraction can be achieved using acidified chloroform/methanol. Efficient extraction of phospholipids from tissues or cell pellets is likely to require additional disruption methods (e.g. homogenization), although essentially complete extraction of monolayers of cells is achieved by brief (1–2 min) exposure to acidified chlorofom/methanol. A procedure for generating and processing inositol phospholipid extracts from cell monolayers is given in *Protocol 4*, which deals with the example of adherent cell monlayers grown in standard multiwell tissue culture plates. The procedure can be adapted for pelletted cells from experiments involving cells in suspension or extraction of whole tissues.

Once the level of Ins(1,4,5)P_3 in each sample has been calculated, a number of corrections must be made to convert these values into original PtdIns(4,5)P_2 mass levels. Although complete hydrolysis of PtdIns(4,5)P_2 is

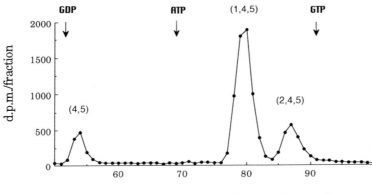

Elution volume (ml)

Figure 2. Separation of the inositol polyphosphate mixture generated by chemical hydrolysis of [³H]PtdIns(4,5)P_2. [³H]PtdIns(4,5)P_2 (~20 000 d.p.m. per sample) was added to chloroform phases obtained after extraction of CHO cell monolayers with acidified chloroform/methanol. Samples were dried under N_2, subjected to hydrolysis with boiling alkali, and neutralized using Dowex-50 (H⁺ form) columns (see *Protocol 4*). The neutral eluate was resolved using the HPLC method detailed in ref. 16. Only data for fractions 50–100 are shown. The peaks were identified relative to GDP/GTP/ATP standards added and by co-elution with authentic [³H]Ins(1,4,5)P_3 and [³H]Ins(4,5)P_2. From four separate experiments it was determined that the relative proportions of Ins(1,4,5)P_3, Ins(2,4,5)P_3 and Ins(4,5)P_2 generated from [³H]PtdIns(4,5)P_2 were 66±2%, 19±3% and 15±3% respectively.

achieved using boiling KOH, Ins(1,4,5)P_3 is not the sole product. In the study of Chilvers *et al.* (15) it was demonstrated that Ins(1,4,5)P_3, Ins(2,4,5)P_3 and Ins(4,5)P_2 were formed in the proportions 66:20:14. Ideally the relative yields of Ins(1,4,5)P_3, Ins(2,4,5)P_3 and Ins(4,5)P_2 should be determined for each biological system studied. This can be achieved by 'spiking' a chloroform cell lipid extract with a small amount (~20 000 d.p.m.) of [³H]PtdIns(4,5)P_2 and processing it through steps 3–6 of *Protocol 4*. The final neutral hydrolysate should be separated by HPLC using a suitable column and gradient elution system (see e.g. ref. 16). An example where the cell lipid extract is derived from CHO cells is shown in *Figure 2*.

Protocol 4. Extraction and hydrolysis of PtdIns(4,5)P_2 for subsequent quantitation using the Ins(1,4,5)P_3 mass assay

Equipment and reagents

- Acidified chloroform/methanol (prepared by mixing 40 volumes chloroform, 80 volumes methanol and 1 volume conc. HCl); chloroform; 0.1 M HCl

- Apparatus for drying down multiple lipid samples (in chloroform) in a stream of N_2, e.g. Techne Sample concentrator (Techne Ltd., Cambridge, UK)

- 1 M TCA; 5% TCA, 1 mM EDTA
- 1 M KOH
- Apparatus for lyophilising multiple 1 ml samples

- Boiling water bath
- Glass wool-plugged columns holding a bed-volume of ~0.5 ml Dowex-50 (200–400 mesh: H^+ form).

Method

1. Following TCA termination of the experiment (see *Protocol 2*), multi-wells are transferred to ice for 15–20 min. The supernatant is then carefully removed without disturbing the cell monolayer, and can be processed for determination of Ins(1,4,5)P_3 or other aqueous-soluble metabolites. The monolayer of cells can then either be processed for lipid extraction directly, or if necessary further washes with 5% TCA/ 1 mM EDTA and distilled water may be performed. (In our hands, the additional wash steps have proved unnecessary for monolayers of cerebellar granule cells, SH-SY5Y neuroblastoma cells, and Chinese hamster ovary cells transfected to express a variety of cell-surface receptors.)

2. Add acidified chloroform/methanol (0.94 ml) to the well and disrupt the cell monolayer by thorough scraping with a plastic pipette tip. After approximately 60 sec take up the medium in the pipette tip and transfer it to a solvent-resistant disposal test tube (5 ml). Longer contact between the acidified chloroform/methanol and the multiwell should be avoided; therefore it is necessary to process each cell monolayer individually.

3. Once all samples have been recovered from the multiwell, add 0.31 ml chloroform and 0.56 ml 0.1 M HCl to each sample, vortex-mix thoroughly and centrifuge the samples (1000 × g, 10 min) to resolve the phases. Carefully aspirate the upper phase and debris at the interface and transfer a known volume (typically 400–450 μl) of the chloroform phase to an Eppendorf tube. Then evaporate the chloroform in a fume-hood in a stream of N_2. At this stage, dry lipid extracts can be purged with N_2, capped and stored at −20 °C for up to 14 days before further processing.

4. For hydrolysis of the lipid fraction to release Ins(1,4,5)P_3, add 0.25 ml 1 M KOH, cap the Eppendorf tubes and place them in a boiling water-bath for 15 min with intermittent vortex-mixing. At the end of the hydrolysis period, transfer the samples to an ice-bath and allow them to cool for 15 min.

5. Neutralize samples by passing them through a column containing Dowex-50 (H^+ form), washing with 5 × 0.25 ml water, and collecting the entire eluate.[a] The pH of the eluate should be checked and adjusted to 6–7 with a small volume of 1 M NaHCO$_3$ if necessary. Before use the column should be washed with 5 ml 1 M HCl and a large volume (> 25 ml) distilled water.

Protocol 4. *Continued*

6. The original description of this procedure (developed for determination of PtdIns(4,5)P_2 mass in bovine tracheal smooth muscle slices, ref. 15) recommended that the eluate be washed with 2 × 2 ml 1-butanol/light petroleum ether (5:1; v/v); however, this has proved unnecessary for the relatively small amounts of tissue (50–100 µg cell protein) routinely involved in cell monolayer experiments.

7. Samples of the eluate can be taken directly for Ins(1,4,5)P_3 mass assay. However, it may prove necessary to lyophilize samples and reconstitute them in a smaller volume of distilled water to give Ins(1,4,5)P_3 concentrations which fall on the central portion of the Ins(1,4,5)P_3 mass assay standard curve.

[a] Loss of Ins(1,4,5)P_3 derived from the hydrolysis of PtdIns(4,5)P_2 can occur at the level of the Dowex column-based neutralization step (15). However, in our experience > 95% of the added Ins(1,4,5)P_3 is collected in the eluate using the elution protocol given above.

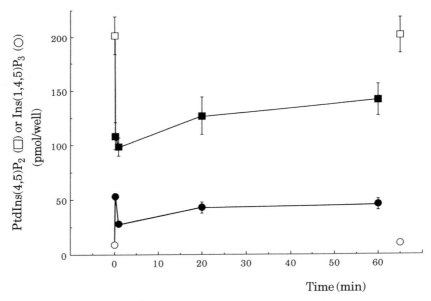

Figure 3 Agonist-stimulated changes in Ins(1,4,5)P_3 and PtdIns(4,5)P_2 mass levels in CHO cells expressing M_3-muscarinic receptors. Monolayers of CHO-m3 cells (transfected to stably express ~1.5 pmol M_3-muscarinic receptor per mg membrane protein) were challenged with 100 µM methacholine (●,■) or vehicle (○,□) for the times indicated. Samples were processed for determination of Ins(1,4,5)P_3 and PtdIns(4,5)P_2 using the methods described in *Protocols 2 and 4*. Data are shown as means ± s.e. mean for four separate experiments performed in duplicate (duplicates pooled for PtdIns(4,5)P_2 determinations). Each well contained 69 ± 6 µg CHO-m3 cell protein.

6. Conclusions

The development of simple methods for the routine assay of the cellular concentrations of PtdIns(4,5)P_2 and Ins(1,4,5)P_3 represents an important advance with respect to studies on the regulation of basal and agonist-stimulated phosphoinositide turnover in general, and specifically to studies concentrating on substrate supply and product generation by the various PLC isozymes present in cells (see *Figure 3* for example). Since its development in 1988, the Ins(1,4,5)P_3 mass assay has become a routine method in signal transduction research, with a number of companies providing (very expensive) assay kits, and variations on the original method emerging (e.g. scintillation proximity assays; see ref. 17). Although the mass assay of PtdIns(4,5)P_2 has been less widely applied to date, this method may prove valuable in delineating the new and emerging roles postulated for PtdIns(4,5)P_2. Furthermore, a variation on the PtdIns(4,5)P_2 extraction/hydrolysis method, coupled with the established Ins(1,3,4,5)P_4 mass assay (13, 18) may provide a method for mass determination of the putative signalling molecule PtdIns(3,4,5)P_3 in the near future.

References

1. Lee, S.B. and Rhee, S.G. (1995). *Curr. Opin. Cell Biol.*, **7**, 183–189.
2. Challiss, R.A.J., Batty, I.H., and Nahorski, S.R. (1988). *Biochem. Biophys. Res. Commun.*, **157**, 684–691.
3. Challiss, R.A.J., Chilvers, E.R., Willcocks, A.L., and Nahorski, S.R. (1990). *Biochem. J.*, **265**, 421–427.
4. Challiss, R.A.J., Willcocks, A.L., Mulloy, B., Potter, B.V.L., and Nahorski, S.R. (1991). *Biochem. J.*, **274**, 861–867.
5. Bredt, D.S., Mourey, R.J., and Snyder, S.H. (1989). *Biochem. Biophys. Res. Commun.*, **159**, 976–982.
6. Challiss, R.A.J. (1995). *Methods in Molecular Biology, Vol 41: Signal transduction protocols*, pp. 167–176, Humana Press Inc., Totowa, NJ.
7. Sharps, E.S. and McCarl, R.L. (1982). *Anal. Biochem.*, **124**, 421–424.
8. Nahorski, S.R. and Potter, B.V.L. (1989). *Trends Pharmacol. Sci.*, **10**, 139–144.
9. Challiss, R.A.J., Jenkinson, S., Mistry, R., Batty, I.H., and Nahorski, S.R. (1993). *Neuroprotocols: A companion to Methods in Neurosciences*, **3**, 135–144.
10. Palmer, S., Hughes, K.T., Lee, K.T., and Wakelam, M.J.O. (1989). *Cell. Signalling*, **1**, 147–153.
11. Willcocks, A.L., Challiss, R.A.J., and Nahorski, S.R. (1990). *Eur. J. Pharmacol.*, **189**, 185–193.
12. Challiss, R.A.J. and Nahorski, S.R. (1993). *Methods Neurosci.*, **18**, 224–244.
13. Donié, F. and Reiser, G. (1989). *FEBS Lett.*, **254**, 155–158.
14. Batty, I.H., Carter, A.N., Challiss, R.A.J., and Hawthorne, J.N. (1997). In *Neurochemistry: a practical approach*, second edition, ed. A.J. Turner and H.S. Bachelard, 229–268, IRL Press, Oxford.

15. Chilvers, E.R., Batty, I.H., Challiss, R.A.J., Barnes, P.J., and Nahorski, S.R. (1991). *Biochem. J.*, **275**, 373–379.
16. Batty, I.H., Letcher, A.J., and Nahorski, S.R. (1989) *Biochem. J.*, **258**, 23–32.
17. Patel, S., Harris, A., O'Beirne, G., and Taylor, C.W. (1995). *Br. J. Pharmacol.*, **115**, 35P.
18. Challiss, R.A.J., and Nahorski, S.R. (1990). *J. Neurochem.*, **54**, 2138–2141.

Desalting inositol polyphosphates by dialysis

JEROEN VAN DER KAAY and PETER J. M. VAN HAASTERT

1. Introduction

In the study of the large spectrum of inositol phosphates and the enzymes involved in their metabolism, anion-exchange chromatography is commonly used for the separation, identification and quantitation of inositol phosphates. Desalting inositol phosphates after chromatography is often desired. The most commonly used method for desalting involves chromatography using a small column packed with mixed ion-exchange resin from which the inositol phosphates are washed with a volatile eluent (e.g. ammonium formate, HCl or Freon/octylamine) which is subsequently removed by vacuum concentration. Other methods used for desalting are gel-filtration or HPLC using a volatile eluent. The large volumes which have to be dried down make this procedure time-consuming. Also, the drying down of small amounts (< 10 nmol) of inositol phosphates can give rise to significant loss of material due to non-specific binding to the tubes. Furthermore, these methods are very laborious when several samples have to be desalted. This section describes the desalting of inositol phosphates by dialysis, which may seem rather peculiar but is very effective and has significant advantages over the above mentioned methods (1).

2. Dialysis

Dialysis is a separation method based on the difference in size of the molecules. The molecular weight cut off (MWCO) of dialysis membranes specifies their pore-size and molecules with a molecular weight (MW) exceeding the MWCO are retained by the membrane, whereas it is generally thought that molecules with a MW below the MWCO can freely diffuse through the pores of the membrane. Dialysis is commonly used for removal of small compounds (salt or nucleotides) from protein or DNA preparations. Surprisingly it can also be applied for the removal of salt from higher inositol phosphates (i.e. from $InsP_3$ up to $[PP]_2InsP_4$), even though the MW of these molecules is 50-fold less than the MWCO of the dialysis membrane. The phenomenon was first encountered during studies on the metabolism of $InsP_6$ in *Dictyostelium* (2). Cell lysates were dialysed in order to remove endogenous $InsP_6$ and other inositol phosphates; the rate and extent of dialysis was monitored by the addition of $[^3H]InsP_6$ to the cell lysates. Unexpectedly, the radiotracer was almost completely recovered in the retentate, although $InsP_6$ has an MW of 924 Da, which is far below the MWCO of 12 000 Da. The Donnan effect, describing the uneven distribution of charged molecules at different sides of a semi-permeable membrane, may provide an explanation for the observed retention of small compounds with multiple charges by dialysis membranes (3).

A series of experiments investigating the principle of retention of inositol phosphates by dialysis membranes is presented below. The method is illustrated with two examples. Finally, the chapter is concluded with a description of the method and its pitfalls that we have observed during the desalting of several hundred samples by dialysis.

3. Dialysis of inositol polyphosphates

Figure 1 shows the retention of inositol, $InsP$, $InsP_3$ and $InsP_6$ when dialysed against 1000 volumes of dialysis buffer for 20 h. Uncharged inositol rapidly diffused through the membrane into the diffusate, and less than 1% remained in the retentate after 20 h. By contrast, about 80% of the $InsP_6$ was recovered in the retentate after overnight dialysis against 1000 volumes of dialysis buffer. $Ins3P$ and $Ins(1,4,5)P_3$ showed intermediate dialysis rates: 25% and 50% respectively were recovered after dialysis for 20 h.

The observed retention is not restricted to inositol polyphosphates since ATP and Glu6P showed essentially the same dialysis characteristics as $Ins(1,4,5)P_3$ and $Ins3P$, respectively (*Figure 2*). Other dialysis membranes with higher MWCO values (10–50 kDa) were tested but showed no significant differences. Adsorption of the radioactive tracer to the dialysis membrane could not explain the observed retention of inositol phosphates since all tested radiotracers showed more than 97% recovery, when retrieved from the solutions outside or inside the dialysis bag. The formation of large complexes exceeding the MWCO does not seem very likely since the extent of dialysis in the presence or absence of protein was essentially identical (data not shown). Dialysis with or without EDTA in the dialysis buffer was also performed, since it is known that inositol phosphates can form complexes with bivalent cations like Mg^{2+} and Ca^{2+} (4). However, the addition of EDTA had no effect on the rate of dialysis (data not shown). Finally, dialysis showed the same kinetics in the presence or absence of 1 mg ml^{-1} phytate hydrolysate, which is a mixture of a large set of unlabelled inositol polyphosphates (see ref. 5) (data not shown).

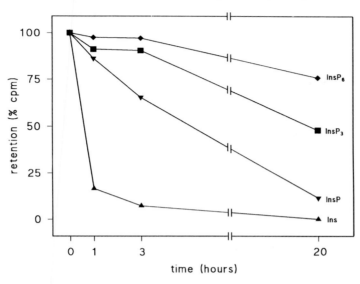

Figure 1. Retention of charged compounds by dialysis. [³H]Inositol, [³H]Ins(3)P, [³H]Ins(1,4,5)P_3 and [³H]InsP_6 were dialysed against 1000 volumes of dialysis buffer (10 mM Hepes, pH 7.1) across a dialysis membrane with a MWCO of 10–12 kDa. Samples of 1 ml contained 0.01 µCi of a radiotracer and 1 mg phytate hydrolysate (5) from which duplicate aliquots of 50 µl were taken at the times indicated. All radioactive tracers were from Amersham, except InsP_6, which was obtained from NEN Dupont. Dialysis membrane with MWCO of 12–14 kDa was from Visking. Dialysis membranes with MWCO values of 10, 15, 25 and 50 kDa were kindly provided by Spectrum Medical Industries Inc.

The dialysis kinetics presented in *Figure 1* seemed inversely related to the charge of the tested compounds. Therefore NaCl was added to the samples and the dialysis buffer to overcome possible charge-effects. The samples used were Glu6P, ATP and InsP_6; in the absence of salt, Glu6P and ATP show the dialysis rate of Ins$P$ and InsP_3, respectively (*Figure 2*). *Figure 2C* reveals that in the presence of high salt concentrations the rate of InsP_6 dialysis was increased, resulting in nearly complete dialysis within 20 h. The effect of 0.1 M and 0.5 M NaCl on the dialysis rate were almost the same, whereas 0.01 M NaCl showed about half maximal effect. The addition of NaCl also increased the diffusion rate of the other charged compounds Ins(1,4,5)P_3/ATP and Ins3P/Glu6P, but had little effect on the dialysis rate of inositol (data not shown).

Finally, *Figure 3* shows the rate of dialysis of samples containing 1 M NaCl when the dilaysis buffer did not contain NaCl. During the first hour, the rate of dialysis of Glu6P, ATP and InsP_6 was relatively fast, but then reduced to a rate approximately equal to that of these compounds in the absence of NaCl in the sample. Apparently, the initial rate of dialysis was mediated by the high NaCl concentration in the sample. Since NaCl also diffuses out of the

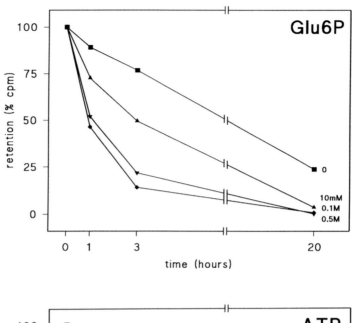

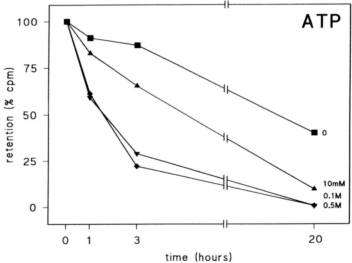

dialysis bag, its concentration rapidly diminishes and becomes too low to effectively shield the charges on the solutes, which are then retained by the dialysis membrane. Thus, the differential dialysis rate of singly charged salt ions and multiply charged inositol polyphosphates suggest the use of dialysis to desalt inositol polyphosphates.

In summary, retention of inositol polyphosphates is solely correlated with

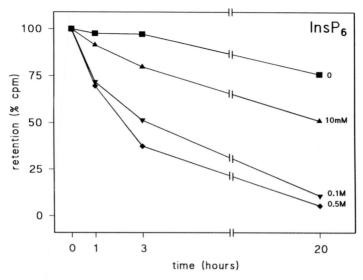

Figure 2. Effect of NaCl on the retention of charged compounds by dialysis. [^{14}C]Glu6P, [^{32}P]ATP, and InsP$_6$ were dialysed against 1000 ml of buffer as described in *Figure 1*. The concentration of NaCl in the dialysis buffer was varied between 0.0 and 0.5 M. Samples contained the same NaCl concentration as the dialysis buffer.

the charge of these molecules and is not restricted to a specific class of compounds. Neither is it due to adsorption of the molecules to the membrane nor to formation of large complexes which exceed the MWCO of the dialysis membrane. The amount of material to be dialysed can be in the millimolar range, and both HPLC analysis and conversion with specific enzymes (6) demonstrate that the compounds do not degrade.

Desalting by dialysis is now routinely performed in our laboratory and has been shown to be successful hundreds of times. The loss resulting from the shielding effect of salt in the samples is a short-time effect and can be minimized by adjusting the time of dialysis so that optimal retention is coupled to effective desalting. With a salt concentration of 1 M, InsP$_3$ is usually dialysed twice for 30 min whilst InsP$_6$ can be dialysed twice for 3 h, which removes over 98% of the salt. The recoveries for InsP$_3$, InsP$_4$, InsP$_5$ and InsP$_6$ can be between 75% and 85% (see *Table 1*).

The main advantages of desalting via dialysis are (i) the method allows many samples to be processed simultaneously, (ii) it is a relatively simple technique, (iii) dialysis can be against any low-ionic-strength buffer or even against distilled water. A disadvantage may be that the method cannot be used for compounds with fewer than two charges, e.g. InsP$_2$. Furthermore desalting by dialysis inevitably leads to the loss of some of the material, depending on the salt concentration and the polarity of the compound (maxi-

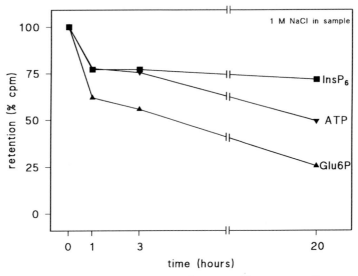

Figure 3. Desalting by dialysis. A mixture of [³H]InsP_6, [³²P]ATP and [¹⁴C]Glu6P containing 1 M NaCl was dialysed against 1000 volumes of dialysis buffer to which no NaCl was added. Samples were take at the times indicated and the radioactivity was determined with a triple label program.

Table 1 Dialysis of inositol phosphates[a]

Compound	Optimal Time	Recovery
Ins(1,4,5)P_3	2 × 30 min	75%
Ins(1,3,4,5)P_4	2 × 1 h	80%
Ins(1,3,4,5,6)P_5	2 × 1.5 h	85%
InsP_6	2 × 3 h	85%

[a]The radioactive compounds were dialysed in the presence of 1 M NaCl against 1 litre of 10 mM HEPES, pH 7.1 for the period indicated; the dialysis buffer was replaced once.

mally 30%). Finally, we have occasionally observed after dialysis and lyophilization that no salt is removed; in those experiments all inositol phosphate was also retained, and a second dialysis always resulted in complete removal of all salt with normal retention of the inositol phosphates.

4. Examples

Two examples are given below in which dialysis was instrumental for the successful removal of small ions. For the preparation of [³²P]Ins(1,3,4,5)P_4 with high specific activity (2000 Ci mmol⁻¹) we phosphorylated unlabelled

Ins(1,4,5)P_3 with [γ-^{32}P]ATP using partially purified recombinant Ins(1,4,5)P_3 3-kinase (6); the purification of [3–^{32}P]Ins(1,3,4,5)P_4 has been optimized. To facilitate the removal of unreacted [γ-^{32}P]ATP, the radioactive phosphate was converted from ATP to glucose 6-phosphate (Glu6P) using hexokinase. The product of the reaction was dialysed for 2 $\times$ 30 min, which removed about 95% of the [^{32}P]Glu6P but retained more than 90% of the [^{32}P]Ins(1,3,4,5)P_4. The contents of the dialysis bag were then purified by HPLC anion-exchanger eluted with a gradient of ammonium phosphate. The fractions containing [^{32}P]Ins(1,3,4,5)P_4 were collected and dialysed for 2 $\times$ 1 h to remove the salt. After lyophilization, re-chromotagraphy of a small sample revealed a purity of more than 99%. Since [3–^{32}P]Ins(1,3,4,5)P_4 has the same high specific activity as its carrier-free [γ-^{32}P]ATP substrate, it is optimally suited for use in an Ins(1,3,4,5)P_4 isotope-dilution assay (7).

A second example demonstrating the usefulness of dialysis is provided by studies on the metabolism of InsP_6 (2). In this study, *Dictyostelium* cells were labelled with 1 mCi [^{32}PO$_4$] for a short period (15 min). Cells were washed, lysed and 99% of the very large excess of [^{32}PO$_4$] (3.7 $\times$ 10^7 cpm) was removed by overnight dialysis against 3 $\times$ 1 litre buffer. Subsequently [^{32}P]InsP_6 (2.4 $\times$ 10^4 c.p.m.) was purified by HPLC anion-exchange chromatography and desalted by overnight dialysis, resulting in a yield of 82% and a purity of 97%. The obtained [^{32}P]InsP_6 was then incubated with *Paramecium* phytase that removed phosphates at a specific order, allowing the determination of the specific activity at each phosphate position in InsP_6. This experiment was instrumental in determining the metabolic route of InsP_6 formation in *Dictyostelium* (2).

Acknowledgement

We thank Peter Van Dijken for many helpful suggestions and for critically reading the manuscript.

References

1. Van der Kaay, J. and Van Haastert, P.J.M. (1995). *Analytical Biochemistry*, **225**, 183.
2. Van der Kaay, J., Wesseling, J., and Van Haastert, P.J.M. (1995). *Biochem. J.*, **312**, 911.
3. Donnan, F.G. (1911). *Z. Electrochem.*, **17**, 572.
4. Luttrel, B.M. (1993). *J. Biol. Chem.*, **268**, 1521.
5. Wregget, K.A., Howe, L.R., Moore, J.P., and Irvine, R.F. (1987). *Biochem. J.*, **245**, 933.
6. Van Dijken, P., Lammers, A.A., Ozaki, S., Potter, B.V.L., Erneux, C., and Van Haastert, P.J.M. (1994). *Eur. J. Biochem.*, **226**, 561.
7. Donie, F., and Reiser, G. (1989). *FEBS Lett.*, **254**, 155.

10

The Ins(1,4,5)P_3 receptor

KATSUHIKO MIKOSHIBA

1. Introduction

Inositol 1,4,5-trisphosphate (Ins(1,4,5)P_3) is the second messenger derived from the hydrolysis of phosphatidylinositol bisphosphate via activation of phospholipase C. The Ins(1,4,5)P_3 binds to its specific receptor, which is an Ins(1,4,5)P_3-induced Ca^{2+}-releasing channel located on the endoplasmic reticulum. In this way the Ins(1,4,5)P_3 signal is converted to a Ca^{2+} signal.

The Ins(1,4,5)P_3 receptor was first characterized in 1979 as a protein called P_{400}, by the studies of lectin binding, partial purification, subcellular distribution, developmental profile and immunocytochemistry (1). This protein was found to be enriched in normal cerebella but virtually absent in cerebella from Purkinje cell-deficient mutant mice (1, 2). Other groups also characterized the protein as PCPP-260 (3), GP-A (4), and GR-250 (5), long before the importance of Ins(1,4,5)P_3 was recognized. In 1988, the receptor protein was purified independently by two groups as an Ins(1,4,5)P_3-binding protein (6) and as a Purkinje cell-enriched protein, P_{400}, which were shown to be identical to Ins(1,4,5)P_3 receptor immunologically (7). Anti-P_{400} monoclonal antibody clearly stained the Ins(1,4,5)P_3 receptor which was highly enriched in the Purkinje cells (8).

There have been many studies on the kinetics of the Ins(1,4,5)P_3-regulated Ca^{2+} channel. It is regulated by kinase A, ATP, GTP, and Ca^{2+}, and has the unique property of quantal Ca^{2+} release, whereby submaximal concentrations of Ins(1,4,5)P_3 cause the partial release of Ca^{2+} from intracellular stores.

Recent molecular cloning studies have revealed that there are at least three types of the Ins(1,4,5)P_3 receptor from distinct genes (9–17,30). Their expressions are differentially regulated among various tissues (18). Although the sequence differences are known, less is understood of the alternative biochemical properties. It is crucially important to characterize each type of receptor biochemically, to explain the tissue-dependent differences in Ca^{2+}-release kinetics. In this chapter, the method for purification and characterization of Ins(1,4,5)P_3 receptor, as well as the method for immunocytochemical localization, are described.

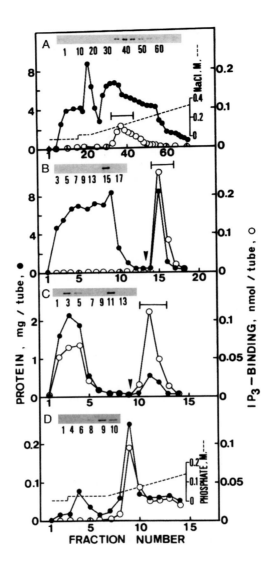

2. Purification of the Ins(1,4,5)P_3 receptor

There are three types of purification methodology, distinguished by the mode of solubilization and the isolation procedure: (I) Solubilization by Triton X-100 or CHAPS (which is a mild detergent) and further purification by various chromatographic columns by using Ins(1,4,5)P_3 binding activity as a marker (6, 19). Ins(1,4,5)P_3-binding assays are described in Chapter 8; (II) Solubilization by the same detergent as (I) and further purification by immunoaffinity using a specific Ins(1,4,5)P_3 receptor antibody (20, 21); (III) solubilization by Zwittergent, which is a vigorous detergent and further purification by various

Figure 1. Purification of type 1 Ins(1,4,5)P3 receptor from the mouse cerebellum (19). The Triton X-100 extract is applied to a DE52 column and the proteins are eluted with a linear NaCl gradient (0.05–0.5 M NaCl indicated by dashed line) (A). The collected fractions indicated by a horizontal bar in (a) are applied to a column of heparin-agarose. After washing the column with a buffer containing 0.25 M NaCl, the receptors are eluted with 0.5 M NaCl (the position of the buffer change is indicated with a arrowhead) (B). The peak fractions of Ins(1,4,5)P_3 binding (indicated by a horizontal bar in (B) are collected and are applied to a column of lentil lectin-Sepharose. After washing, the receptors are eluted with 0.8 M alpha-methyl-D-mannoside (the position of the application of alpha-methyl-D-mannoside is indicated by a arrowhead) (C). The receptor fractions indicated by horizontal bar in (C) are applied to a column of hydroxyapatite. The receptors are eluted with a linear sodium phosphate gradient (0.025–0.15 M sodium phosphate indicated by dashed line) (D). Aliquots of the each fraction were assayed for protein; results of the immunoblotting of the fractions with a monoclonal antibody against Ins(1,4,5)P_3 receptor are also indicated.

chromatographic columns detecting the Ins(1,4,5)P_3 receptor by Western blotting (7).

2.1 Purification method (I)

The procedure utilizes Triton X-l00 or CHAPS to solubilize the receptor in the membrane using Ins(1,4,5)P_3 binding as a marker. The Ins(1,4,5)P_3 receptor purified by this procedure has a tetrameric structure. This is a standard purification procedure to isolate native receptors (19) (*Figure 1*).

Protocol 1. Purification of the Ins(1,4,5)P_3 receptor from mouse cerebellum (I)

Reagents

- [^{3}H]Ins(1,4,5)P_3 (NEN)
- Ins(1,4,5)P_3 and heparin agarose (Sigma)
- Bio Gel HTP (Bio Rad)
- Buffer A: 1% Triton X-100, 1 mM EDTA, 10 μM leupeptin, 1 mM 2-mercaptoethanol, 50 mM Tris-HCl, pH 8.0
- Buffer B: 0.2% Triton X-100, 10 μM leupeptin, 0.1 mM phenylmethylsulfonyl fluoride (PMSF), 1 mM 2-mercaptoethanol, 50 mM Tris-HCl, pH 8.0

- Buffer C: 0.5 M NaCl, 0.2% Triton X-100. 10 μM leupeptin, 10 μM pepstatin A, 0.1 mM PMSF, 20 mM Tris-HCl, pH 8.0, 1 mM 2-mercaptoethanol
- Buffer D: 0.2% Triton X-100, 0.5 M NaCl, 1 mM 2-mercaptoethanol, 25 mM sodium phosphate, pH 8.0
- Lentil lectin-Sepharose (Pharmacia)
- DE52 (Whatman)

A. Membrane preparation[a]

1. Anesthetize adult ddY mice and then kill by decapitation and dissect the cerebella. Perform all the procedures described below at 0–4°C.

2. To the 10 grams of fresh cerebella , add 9 vol of the ice-cold solution containing 0.32 M sucrose, 1 mM EDTA, 0.1 mM PMSF 10 μM leupeptin, 10 μM pepstatin A, 1 mM 2-mercaptoethanol and 5 mM Tris-HCl, pH 7.4.

3. Homogenize in a glass-Teflon Potter homogenizer with 10 strokes at 850 r.p.m.

Protocol 1. *Continued*

4. Centrifuge the homogenate at 1000 × *g* for 5 min at 2°C, and wash it again under the same conditions.

5. Combine the supernatants and centrifuge at 105 000 × *g* for 60 min at 2°C to precipitate P2 (crude mitochondrial) + P3 (microsomal) fractions.

B. *Solubilization and purification of Ins(1,4,5)P₃ receptor protein*

1. Suspend either the P31 fraction (see *Protocol 1A, footnote a*) or the P2 + P3 fraction (see *Protocol 1*, step 5) in the solution containing 1 mM EDTA, 0.1 mM PMSF, 10 μM leupeptin, 10 μM pepstatin A, 1 mM 2-mercaptoethanol and 50 mM Tris-HCl, pH 8.0.

2. Solubilize the membrane by addition of 10% (w/v) Triton X-100 to give final protein and detergent concentrations of 3.0 mg ml⁻¹ and 1.0%, respectively.

3. Stir the sample for 30 min at 0°C and then centrifuge at 20 000 × *g* for 60 min at 2°C. Apply the supernatant to a column of DE52 (2.6 × 10 cm) equilibrated in buffer A.

4. Wash the column with 20 ml of 0.05 M NaCl/buffer A, and then elute the receptor with a 400 ml linear NaCl gradient (0.05–0.5 M in buffer A).

5. Pool the peak fractions containing Ins(1,4,5)P₃ binding activities and apply to a column of heparin-agarose (0.97 × 7 cm) equilibrated in buffer B containing 0.25 M NaCl.

6. Wash the column with 30 ml of 0.25 M NaCl/buffer B and then elute the receptor with 30 ml of 0.5 M NaCl/buffer B.

7. Pool the peak fraction of the Ins(1,4,5)P₃ binding activities and apply to a column of lentil lectin-Sepharose (0.95 × 2.8 cm) equilibrated in buffer C.

8. Wash the column with 8 ml of buffer C and elute the receptors with 0.8 M α-methyl-D-mannoside/buffer C.

9. Pool the peak fractions of the Ins(1,4,5)P₃ binding activity and absorb the proteins to a column of Bio Gel HTP (0.7 × 0.8 cm) equilibrated in buffer D.

10. Wash the column with 2 ml of buffer D and elute the receptors with a 10 ml linear sodium phosphate gradient (25–150 mM).

[a]There is an alternative procedure to isolate highly enriched Ins(1,4,5)P₃ receptor fraction. Instead of steps 4 and 5 in *Protocol 1A* above, the homogenate is processed as follow:
(i) Centrifuge the suspension at 12 000 × *g* for 15 min at 2°C, and wash again under the same conditions.
(ii) Overlay the combined supernatants which contain microsomes (P3) over a 0.8 M sucrose solution and centrifuge at 52 800 × *g* for 2 h at 2°C.
(iii) Recover the band at the interface between 0.32 M and 0.8 M sucrose, and dilute with cold distilled water and centrifuge at 105 000 × *g* for 1 h at 2°C to yield the so-called P3l fraction.
The protein recovery of Ins(1,4,5)P₃ receptor in the P31 fraction is low but the enrichment ratio of the receptor is very high.

2.2 Purification method (II)

This procedure uses Triton-X100 or CHAPS to solubilize the receptor, as described above for method I, but a type-specific antibody is used to purify the receptor (*Protocol 2*). This procedure needs specific antibody, but requires less time than other methods. This is a suitable way to analyse the channel activity since purification requires considerably less time so denaturation of the receptor is minimized (20, 21). *Table 1* shows the elution profile for step 6 of *Protocol 2B*.

Protocol 2. Purification of the Ins(1,4,5)P_3 receptor from mouse cerebellum (II)

Reagents

- Phosphatidylcholine (Avanti Polar Lipids)
- CNBr activated Sepharose 4B (Pharmacia LKB Biotechnology Inc.)
- Buffer A: I mM EDTA, 1 mM 2-mercaptoethanol, 0.1 mM PMSF, I0 μM leupeptin, 10 μM pepstatin A, and 50 mM Tris-HCl, pH 7.4

- Buffer B: Buffer A, and 5 μM pep 6 peptide, 1 μg ml⁻¹ phosphatidyl choline
- Pep 6: GHPPHMNVNPQQ, C-terminal peptide of the type 1 mouse Ins(1,4,5)P_3 receptor (18, 19, 20)

A. *Preparation of the immuno-affinity column (an anti-pep 6 antibody-Sepharose 4B beads)*

1. Purify the immunoglobulin G (IgG) fraction of the anti/serum by ammonium sulfate precipitation and dialyse it against 0.1 M NaHCO₃ and 0.5 M NaCl, pH 8.3, and then couple 10 mg of IgG to 1 ml of CNBr-activated Sepharose 4B according to the manufacturer's protocol.

B. *Membrane solubilization*

1. Resuspend the P2 + P3 membrane fraction (*see Protocol 1*) in buffer A.
2. Add 10% CHAPS to give final protein and detergent concentration of 3.0 mg ml⁻¹ and 1.0%, respectively, and stir for 30 min at 0°C.
3. Centrifuge the solution at 20 000 × *g* for 60 min at 2°C and save the supernatant.
4. Incubate the supernatant with the anti-pep 6 antibody-Sepharose 4B beads for 2 h at 4°C under gentle stirring.
5. Transfer the beads into a column and wash with 50 ml of buffer A containing 1 mg of phosphatidylcholine per ml.
6. Elute the Ins(1,4,5)P_3 receptor by a batch method. Add 1 ml of buffer B / 1 ml of packed beads and stir gently for 20 min at 4°C, and then centrifuge the beads at 100 × *g* for 30 sec. Repeat the elution step at least 5 times. Regenerate the beads with 1 M glycine and 1.5 M NaCl, pH 2.5.

Table 1. Immunoaffinity purification of Ins(1,4,5)P_3 receptor from mouse cerebellum[a](20)

Sample	Total [³H]Ins(1,4,5)P_3 binding sites (pmol)	Yield of [³H]Ins(1,4,5)P_3 binding sites (%)
Starting material	1933 ± 241	100
First cycle		
Flow-through	1103 ± 346	57.1
Wash	250 ± 88	12.9
Eluate 1	22 ± 14	1.1
Eluate 2	33 ± 11	1.7
Eluate 3	43 ± 11	2.2
Eluate 4	39 ± 9	2.0
Eluate 5	35 ± 10	1.8
Second cycle		
Flow-through	287 ± 93	14.9
Wash	77 ± 10	4.0
Eluate 6	3.9 ± 0.6	0.2
Eluate 7	8.4 ± 2	0.4
Eluate 8	12 ± 3	0.6
Eluate 9	13 ± 4	0.7
Eluate 10	14 ± 7	0.7
Third cycle		
Flow-through	64 ± 10	3.3

[a]Ins(1,4,5)P_3 receptor is solubilized from 16 mice cerebella with buffer A containing 1% CHAPS and incubated with antibody-Sepharose 4B beads for 2 h at 4°C under gentle stirring as described in *Protocol 2*. Ins(1,4,5)P_3 receptor is eluted, and the flow-through fraction is reincubated with the regenerated beads to maximize binding of Ins(1,4,5)P_3 receptor (cylcles 1–3). Specific elution is performed by C-terminus antibody at a concentration of 5 μM in the first and second cycles. Specific [³H]Ins(1,4,5)P_3 binding activity is assayed by the centrifugation method. The values are the means ±S.D. for 4–8 experiments.

2.3 Purification method (III)

This method was originally developed by Maeda et al. (7), and utilizes Zwittergent for solubilization of the receptor, using specific antibody against Ins(1,4,5)P_3 receptor as marker (*Protocol 3*).

Protocol 3. Purification of Ins(1,4,5)P_3 receptor from the mouse cerebellum (III)

Reagents

- Sepharose CL-4B and Con A-Sepharose (Pharmacia)
- Polyethylene glycol 1500 (Boehringer Mannheim)
- Zwittergent 3–14 (Calbiochem)
- Con A (EY Laboratories)

- Buffer A: 10 mM EDTA, 10 mM *N*-ethyl-maleimide (NEM), 1 mM PMSF, 0.036 mM pepstatin A, 50 mM Tris-HC1, pH 8.0
- Buffer B: 0.5% Triton X-100, 0.5% Zwittergent 3–14, 0.5 M NaCl, 1 mM CaCl₂, 1 mM MnCl₂, 20 mM Tris-HCl, pH 7.0

A. *Solubilization of the receptor*

1. Add the P31 fraction obtained from 50 grams of mouse cerebella (see *Protocol 1*, step 6) to 100 ml of buffer A containing 2% Triton X-100, and stir the suspension for 1 h at 0°C.

2. Centrifuge the suspension at 105 000 × *g* for 1 h at 2°C.

3. Treat the pellet with 25 ml of buffer A containing 8% Zwittergent 3–14 for 10 min at 0°C.

4. Add 25 ml of buffer A containing 8 M guanidinium chloride and stir the solution overnight at 0°C, and centrifuge at 105 000 × *g* for 1 h at 2°C.

5. Apply the extract to the Sepharose CL-4B column (5 × 100 cm) and elute the proteins at a flow rate of 80 ml h⁻¹ with 4 M guanidinium chloride, 0.5% Zwittergent 3–14, 0.5% Triton X-100, 5 mM EDTA, 20 mM Tris-HCl, pH 8.0.

6. Collect 13 ml for each fraction and assay aliquots by sodium dodecyl sulfate-polyacrylamide gel electrophoresis (SDS-PAGE) as follow

   (i) Before SDS-PAGE, mix the samples with three volumes of 95% ethanol containing 1.3% potassium acetate to remove guanidinium chloride (ethanol precipitation).

   (ii) After centrifugation, dissolve the resultant precipitates in 2% SDS/5% 2-mercaptoethanol/0.0625 M Tris-HCl, pH 6.8/10% glycerol (SDS-PAGE sample buffer) by heating in a boiling water bath for 10 min.

   (iii) Detect Ins(1,4,5)P₃ receptor by carrying out SDS-PAGE (linear 5–12.5% or 5% polyacrylamide slab gel) by the method of Laemmli (22).

B. *Con-A Sepharose chromatography*

1. Concentrate the combined fractions from Sepharose CL-4B chromatography with an Amicon PM30 membrane to 3 ml, and dialyse five times against 30 volumes of 0.5% Triton X-100, 0.5% Zwittergent 3–14, 0.5 M NaCl, 20 mM Tris-HCl, pH 7.0.

2. Apply the dialysate to a Con A-Sepharose column (0.87 × 17 cm), which has been equilibrated with buffer B. Maintain the flow rate at 10 ml h⁻¹.

3. Wash the column with buffer A, and then elute the absorbed proteins with 0.5 M - methyl-D-mannoside/buffer A, and 0.5% Triton X-100, 0.5% Zwittergent 3–14, 0.1 M acetic acid.

4. Neutralize the fractions which was eluted with acetic acid with 1.0 M Tris.

Protocol 3. *Continued*

5. Collect the fractions and add solid guanidinium chloride and 250 mM EDTA to give the final concentrations of 2 M and 5 mM, respectively.

6. Apply the samples to a column of Sepharose CL-4B.

C. *Rechromatography on a Sepharose CL-4B column*

1. Concentrate the samples with an Amicon PM30 membrane to 3 ml, and apply to a Sepharose CL-4B column (1.6 × 100 cm). The eluate is the same as that used for the Sepharose CL-4B chromatography described above, but the flow rate is 10 ml h^{-1}.

2. Collect the fractions, and concentrate with an Amicon PM30 membrane, and store at $-80\,°C$.

3. Functional assay of Ins(1,4,5)P_3 receptor

3.1 Measurements of Ins(1,4,5)P_3-induced Ca^{2+} release by the immunoaffinity-purified Ins(1,4,5)P_3 receptor

There are many reports characterizing the properties of Ins(1,4,5)P_3-regulated Ca^{2+}-release channel using permeabilized cells and microsomal preparations, which have been complicated by the following factors:

(i) Composition of subtypes of Ins(1,4,5)P_3 receptors. The presence of multiple Ins(1,4,5)P_3 receptor types in single cells may affect the kinetics of Ca^{2+} release.

(ii) Metabolism of Ins(1,4,5)P_3. Ins(1,4,5)P_3 could easily be metabolized by specific or non specific kinases and phosphatases which may be present in crude systems. The concentration of ligand during experiments is known to be one of the critical factors. Most of the properties of Ins(1,4,5)P_3-regulated Ca^{2+} release (multiple affinity sites on single Ins(1,4,5)P_3 receptors, quantal release by submaximal doses, inactivation by Ins(1,4,5)P_3 itself) are dependent on Ins(1,4,5)P_3 concentration.

(iii) Ca^{2+} pump. The activity of the ATP-driven Ca^{2+} pump affects Ins(1,4,5)P_3 induced Ca^{2+} release (IICR) by refilling Ca^{2+} stores. This prevents us from evaluating the cooperativity of IICR by reducing the IICR net to a greater extent at low concentrations of Ins(1,4,5)P_3 than at high concentrations.

(iv) Molecules sensing changes in Ca^{2+} concentration. Dynamic changes in cytosolic and luminal Ca^{2+} concentrations have been argued to be involved in functional regulation of Ca^{2+} release properties by modifying the function of the Ins(1,4,5)P_3 receptor itself and by activating Ins(1,4,5)P_3 receptor-modulator proteins, e.g. protein kinases such as Ca^{2+}-calmodulin dependent protein kinase and protein kinase C, phosphatases such as calcineurin, Ca^{2+} binding proteins such as calmedin, and $Ins(1,4,5)P_3$-metabolizing enzymes such as $Ins(1,4,5)P_3$ kinase.

(v) Heterogeneity in $Ins(1,4,5)P_3$ sensitive pools. There is a subcellular heterogeneity in $Ins(1,4,5)P_3$ sensitive Ca^{2+} stores, e.g. subsurface cisternae, calciosomes, nuclear membranes, etc. which may have different Ca^{2+} release properties. Artificial effects on $Ins(1,4,5)P_3$ sensitive Ca^{2+} stores by experimental conditions must be considered, e.g. fusion of cisternae membranes by excess treatment with saponin and induction of formation of cisternal stacks mediated by $Ins(1,4,5)P_3$ receptors by non-physiological treatment (23, 24).

Structural and functional studies using various deletion mutants of the type 1 receptor revealed that the receptor is composed of three functional domains. An $Ins(1,4,5)P_3$ binding domain is located at the N-terminal, a channel domain which is composed of six membrane spanning regions is located at the C-terminal, and a regulatory domain which couples the $Ins(1,4,5)P_3$ receptor with other signal transduction systems is located at the central region that links both $Ins(1,4,5)P_3$ binding and channel domain. Function of the $Ins(1,4,5)P_3$ receptor can be studied by $Ins(1,4,5)P_3$ binding and $Ins(1,4,5)P_3$ induced Ca^{2+} release assays. Ca^{2+} releasing activity is studied using the $Ins(1,4,5)P_3$ receptor reconstituted in liposome (see *Figure 2*). Cross linking experiments clearly showed that $Ins(1,4,5)P_3$ receptor forms tetrameric structure (20, 25). Here, studies on Ca^{2+} release by purified $Ins(1,4,5)P_3$ receptor reconstituted into liposome, as well as the tetrameric structural analysis, are described.

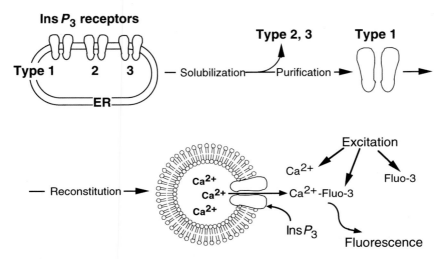

Figure 2. Schematic diagram of the incorporation of purified Ins(1,4,5)P₃ receptor into proteoliposomes. (Figure by K. Hirota).

Protocol 4. Reconstitution of the immunoaffinity-purified Ins(1,4,5)P_3 receptor into liposomes

Reagents

- Centriprep 100 (Amicon)
- Ins(1,4,5)P_3, CHAPS, and fluo-3 are from Dojindo Laboratories (Kumamoto, Japan)
- Phosphatidylcholine, phosphatidylserine, and cholesterol (Avanti Polar Lipids, Inc.)

- Chelex 100 (Bio-Rad)
- Buffer A: 100 mM KCl, 1 mM 2-mercaptoethanol, 10 mM HEPES-KOH [pH 7.4] and 4 mM $CaCl_2$

Method

1. Dissolve phosphatidylcholine, phosphatidylserine and cholesterol in chloroform to give a concentration of 3.0, 1.0 and 0.8 mg ml^{-1}, respectively.

2. Dry the lipid mixture to a thin film under a stream of nitrogen gas and then under vacuum.

3. Suspend the lipid film at 2 mg ml^{-1} in buffer A containing 1% CHAPS.

4. Concentrate the immunoaffinity-purified Ins(1,4,5)P_3 receptor by using Centriprep 100 (Amicon) to give a protein concentration of ~100 μg ml^{-1}.

5. Mix the concentrated Ins(1,4,5)P_3 receptor solution with buffer A containing lipids and detergent to give the following final concentrations: 50 μg ml^{-1} Ins(1,4,5)P_3 receptor, 0.5 mg ml^{-1} lipids and 1% CHAPS.

6. Incubate the mixture for 20 min on ice with occasional gentle stirring.

7. Dialyse the Ins(1,4,5)P_3 receptor-lipid mixtures for 72 h against 8 changes of a 500- fold volume excess of buffer A at 4°C.

8. Centrifuge the solution at 100 000 × g for 30 min at 2°C to pellet the proteoliposomes (Ins(1,4,5)P_3 receptor in lipid vesicles) .

9. Wash the buffer with buffer B (buffer A without Ca^{2+} + 10 or 1 μM of fluo-3) twice, and resuspend with buffer B in the same volume used before dialysis.

10. After incubation for 10 min at 25°C, pass the resuspended proteoliposomes over Chelex-100 to remove Ca^{2+}. This is used for the Ca^{2+} release assay.

3.2 Removal of Ca^{2+} contamination

Removal of Ca^{2+} contamination is necessary to measure the calcium release and to improve the sensitivity of the fluorometric measurements. We remove the Ca^{2+} contamination according to the method of Meyer *et al.* (26) (*Protocol 5*).

Protocol 5.   Removal of Ca^{2+} contamination

Reagents

- Diethylenetriamine-N,N,N',N'', N''-penta-acetic acid-conjugated polymetal sponge (Molecular Probes)
- Ins(1,4,5)P_3 [^{3}H] radioreceptor assay kit (NEN Research Products)

Method

1. Pass all solutions to be used in fluorometric measurements over a polymetal sponge and wash all labwares successively with detergent, 0.1 N HCl, distilled water and the buffer to be used.

2. Check Ca^{2+} contamination in all solutions, cuvettes and stir bars using the Ca^{2+} indicator fluo-3 before the measurements.

3. Pass also the Ins(1,4,5)P_3 stock solution over the polymetal sponge to remove Ca^{2+}. Check whether this process does not cause any changes in Ins(1,4,5)P_3 concentrations by using [^{3}H] Ins(1,4,5)P_3 radio receptor assay kit.

3.3 Ins(1,4,5)P_3 induced Ca^{2+} release measurements

Ca^{2+} efflux from the proteoliposomes is measured by monitoring the fluorescence changes of fluo-3 (*Protocol 6*). Fluorometric measurements of IICR are performed by using an F-2000 fluorometer (Hitachi, INC.) interfaced to a PC9801-VX computer (NEC, Inc.). The excitation and emission wave length are 500 and 525 nm respectively with 10 nm bandpass. Fluorescence signals are corrected for fluctuations in excitation light intensity.

Protocol 6.   Ins(1,4,5)P₃ induced Ca^{2+} release measurements

Equipment

- Fluorometer F-2000 (Hitachi, Inc.) interfaced to a PC 9801-VX computer (NEC, Inc.)

Method

1. Put the 0.4 ml of the proteoliposome solution in a 0.5 × 0.5 cm quartz cuvette at 25°C with continuous-stirring by a Teflon stir bar.

2. Monitor the Ca^{2+} release after addition of 2 μl Ins(1,4,5)P_3 to give the desired Ins(1,4,5)P_3 concentration. Obtain data every 200 ms.

3. Calibrate the fluorescent intensities of fluo-3 to free Ca^{2+} concentrations using Ca^{2+}–EGTA buffering system. The calibration curve gives the dissociation constant of fluo-3 for Ca^{2+} as 170 nM, which is used to estimate the free and total Ca^{2+} concentrations.

Protocol 6. *Continued*

To exclude the possibility of Ca^{2+} regulation of Ca^{2+} release, we use 10 μM fluo-3, the concentration of which is high enough to buffer the released Ca^{2+} and to keep deviations of extravesicular free Ca^{2+} concentration within 10 to 30 nM.

We also examined Ca^{2+} release using 1 μM fluo-3, where the deviations of free Ca^{2+} concentration were 150–300 nM, to compare the effects of changes in free Ca^{2+} concentration on Ins(1,4,5)P_3-mediated Ca^{2+} release.

3.3.1 Time Course of Ca^{2+} release by the immunoaffinity-purified Ins(1,4,5)P_3 receptor

Figure 3 shows a typical profile of Ca^{2+} release by the immunoaffinity-purified Ins(1,4,5)P_3 receptor reconstituted into lipid vesicles. Five hundred nM Ins(1,4,5)P_3 induced Ca^{2+} release from the liposomes followed a constant leakage of Ca^{2+} (*Figure 4A*), which is linear over the time range of the experiments. The rate of leak from the liposomes is calculated to be about 1.5 nM sec^{-1}. The net Ca^{2+} release (*Figure 4B*) is obtained by extrapolating and subtracting the constant Ca^{2+} leakage (*Figure 4A*, solid line) from the profile. The net Ca^{2+} release cannot be fitted by a single exponential but is found to be a biexponential (*Equation 1*) (*Figure 4C*, solid line) with the fast and slow rate constants (k_{fast} = 0.51±0.01 sec^{-1} (71±1%), k_{slow} = 0.042±0.001 sec^{-1} (29±1%)), indicating that the purified Ins(1,4,5)P_3 receptor has two states for Ins(1,4,5)P_3-mediated Ca^{2+} release.

$$\Delta[Ca^{2+}]_{total} = T\,(1 - A_{fast} \bullet e^{-kfast \bullet t} - A_{slow} \bullet e^{-kslow \bullet t}) \tag{1}$$

where T represents a total amount of released Ca^{2+}, A is amplitude of the fast and slow components (%) ($A_{fast} + A_{slow}$ = 100%), k is the rate constant (sec^{-1}) and t is time (sec).

3.3.2 Kinetic analysis of Ins(1,4,5)P_3-mediated Ca^{2+} release

Different concentrations of Ins(1,4,5)P_3 are added to obtain dose-response curves. *Figure 3* shows typical time courses of Ca^{2+} release observed using the same batch of proteoliposomes. Submaximal concentrations of Ins(1,4,5)P_3 causes partial Ca^{2+} releases, and rates of Ca^{2+} release are dependent on the Ins(1,4,5)P_3 concentration.

Protocol 7. Cross-linking of denatured and nondenatured Ins(1,4,5)P_3 receptor

Reagents

- Ethylene glycol-bis (sulfosuccinimidyl succinate) (S-EGS) (Pierce Chemical Co.)
- Buffer A: 1 mM EDTA, 1 mM 2-mercapto ethanol, 0.1 mM PMSF, 10 μM leupeptin, 10 μM pepstatin A, and 50 mM Tris-HCl, pH 7.4

- Buffer B: phosphate-buffered saline (PBS), 1% CHAPS, 0.1 mM PMSF, 10 μM leupeptin, and 10 μM pepstatin A

Method

1. Denature the immunoaffinity-purified Ins(1,4,5)P_3 receptor with 6 M urea, 20 mM dithiothreitol, and 1% SDS in buffer A containing 1% CHAPS for 5 min at 95°C.

2. Dialyse against buffer B to remove Tris, urea, and dithiothreitol.

3. Adjust the concentration of SDS to 0.3% by dilution of the sample with PBS containing 0.1 mM PMSF, 10 μM leupeptin, and 10 μM pepstatin A.

4. Treat aliquots (50 μl) of each solution with 10 mM S-EGS, cross linker.

5. Allow the reactions to proceed for 30 min at 0°C before quenching by the addition of 10 μl of 50 mM Tris-HCl, pH 8.0.

6. Solubilize 10 μl of sample in agarose-PAGE buffer at a final concentration of 1% SDS, 1 mM EDTA, 5% 2-mercaptoethanol, 10 mM Tris-HCl, pH 8.0, 10% glycerol and heat in a boiling water and bath for 3 min.

7. Analyse the sample with agarose-PAGE (27) and immunoblotting.

[a]Steps 1 and 2 are for denaturing condition. For nondenatured receptor, start from step 3.

Protocol 8. Cross-linking of proteoliposomes reconstituted with Ins(1,4,5)P_3 receptor to demonstrate the tetrameric structure[a]

Method

1. Centrifuge 100 μl of the proteoliposomes at 100 000 × g for 20 min at 2°C.

2. Resuspend the resulting pellet in 50 μl of 50 mM sodium phosphate, pH 8.0.

3. Treat the suspension with various concentrations of cross-linker, S-EGS.

Protocol 8. *Continued*

4. Allow the reactions to proceed for 30 min at 0°C and then add 10 μl of 50 mM Tris-HCl, pH 8.0 for quenching.

5. Solubilize 10 μl of sample in agarose-PAGE (27) sample buffer at a final concentration of 1% SDS, 1 mM EDTA, 5% 2-mercaptoethanol, 10 mM Tris-HCl, pH 8.0, 10% glycerol. Heat in a boiling water bath for 3 min.

6. Analyse the samples with agarose-PAGE and immunoblotting.

[a]See ref. 20.

4. Analysis of the Ins(1,4,5)P_3 receptor family

The Ins(1,4,5)P_3 receptor exists as a tetrameric complex to form a functional Ins(1,4,5)P_3-gated Ca^{2+} channel. Molecular cloning studies have shown that there are at least three types of Ins(1,4,5)P_3 receptor subunits, designated type 1, type 2, and type 3. The levels of expression of Ins(1,4,5)P_3 receptor subunits in various cell lines are investigated by Western blot analysis using type-specific antibodies against 15 C-terminal amino acids of each Ins(1,4,5)P_3 receptor subunit. We found that all the three types of Ins(1,4,5)P_3 receptor subunits were expressed in each cell line examined, but their levels of expression varied. To determine whether Ins(1,4,5)P_3 receptors form heterotetramers, we employed immunoprecipitation experiments using Chinese hamster ovary cells (CHO-K1 cells), in which all three types are abundantly expressed. Each type-specific antibody immunoprecipitated not only the respective cognate type but also the other two types (28). This result suggests that distinct types of Ins(1,4,5)P_3 receptor subunits assemble to form heterotetramers in rat liver, in which Ins(1,4,5)P_3 receptor type 1 and type 2 are expressed abundantly. Previous studies have shown some functional differences among Ins(1,4,5)P_3 receptor types, suggesting the possibility that various compositions of subunits confer distinct channel properties. The diversity of Ins(1,4,5)P_3 receptor channels may be further increased by the co-assembly of different Ins(1,4,5)P_3 receptor subunits to form homeo- or heterotetramers.

4.1 Preparation of type-specific antibodies against each type of the Ins(1,4,5)P_3 receptor

In order to characterize the biochemical properties and localization of the three types of Ins(1,4,5)P_3 receptors it is essential to obtain type-specific antibody against each type of receptor. Fortunately, the C-terminal of each receptor differs from receptor to receptor. Detailed description of the raising of antibodies are described in *Protocol 9.*

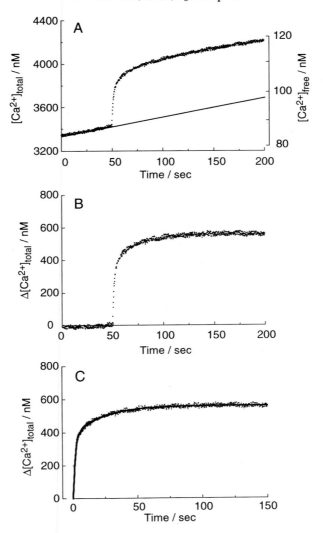

Figure 3. Ins(1,4,5)P_3-induced Ca^{2+} release profile from proteoliposomes reconstituted with the purified Ins(1,4,5)P_3 receptor (21). Changes of fluorescence of the Ca^{2+} indicator fluo-3 ([fluo-3]=10 μM) are recorded after injection of Ins(1,4,5)P_3 (500 nM). The total Ca^{2+} concentration is estimated from the fluorescent intensity as described in the method. (A) Ins(1,4,5)P_3-induced Ca^{2+} release from the liposomes is followed by a constant leakage of Ca^{2+} (the solid line). (B) The net IICR is obtained by extrapolating and subtracting the constant Ca^{2+} leakage from the profile. (C) The net IICR is found to be well fitted by a biexponential (the solid line) with the fast and slow rate constants.

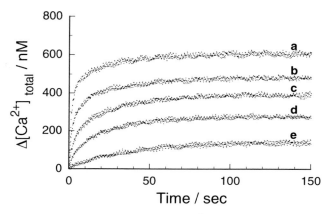

Figure 4. Time course of Ins(1,4,5)P_3-induced Ca^{2+} release following the injection of different Ins(1,4,5)P_3 concentrations (21). Ins(1,4,5)P_3-induced Ca^{2+} release at different concentrations of Ins(1,4,5)P_3 is performed on a single batch of the proteoliposomes ([fluo-3] = 10 μM). 5 μM (a), 200 nM (b), 70 nM (c), 40 nM (d) and 20 nM (e) of Ins(1,4,5)P_3.

Protocol 9. Preparation of peptide antibodies

Reagents

- GHPPHMNVNPQQ (pep 6, amino acid residues 2736–2747 of mouse type1 Ins(1,4,5)P_3 receptor)
- CLGHPPHMNVNPOOPA (peptide-IC, amino acid residues 2681–2695 of human type 1 Ins(1,4,5)P_3 receptor)
- CLGSNTPHVNHHMPPH (peptide-2C, 2687–2701 of human type 2 Ins(1,4,5)P_3 receptor)
- CRQRLGFVDVQNS*ISR (peptide-3C, 2657–2671 of human type 3 Ins(1,4,5)P_3 receptor)
- CGESLANDFLYSDVS*R (peptide-1L, 2483–2497 of human type 1 Ins(1,4,5)P_3 receptor)
- CGSHQVPTMTLTTMME (peptide-2L, 2436–2450 of human type 2 Ins(1,4,5)P_3 receptor)
- CSPLGMPHGAAAFVDT (peptide-3L, 2410–2424 of human type 3 Ins(1,4,5)P_3 receptor)

Serines designated as S* are introduced instead of the original cysteins to avoid disulfide bond formation. Cysteins are attached to all of the peptides as the N-terminal spacer amino acids used as the coupling sites for the preparation of keyhole limpet hemocyanin (KLH; Calbiochem, USA)-peptide conjugates.

The peptide is conjugated to bovine serum albumin (BSA) via 1-ethyl-3-(3-dimethylaminopropyl) carbodiimide (EDC) as described by Richardson *et al.* (29).

- Sephadex G-50 (Pharmacia)
- Freund's complete adjuvant, Freund's incomplete adjuvant (Sigma)

Method

1. Dissolve 20 mg of peptides and 20 mg of BSA in 5 ml of PBS, pH 7.3. Add 20 mg of EDC at 4°C with constant stirring and stir the mixture overnight.

2. Separate the remaining unreacted EDC and peptides by gel filtration on a Sephadex G-50 column equilibrated in 50 mM ammonium acetate.

3. Collect the fractions containing peptide-BSA, lyophilize, and dissolve them in PBS.

4. Immunize New Zealand White rabbits by intradermal injection with a homogenate containing 1 ml of Freund's complete adjuvant and 1 ml of peptide-BSA conjugate (200 μg of peptide).

5. Three weeks later, perform booster injections into the rabbits using the homogenate containing 1 ml of Freund's incomplete adjuvant and 1 ml of the conjugate (100 μg of peptide).

6. Collect antiserum each week thereafter.

7. Perform booster injections every 2 weeks until the titer of the antiserum is saturated.

Protocol 10.   Antibody purification by affinity chromatography

Reagents
- Affi-Gel 10 (Bio-Rad)
- Protein A-Sepharose CL-4B (Pharmacia)

Method

1. Couple the peptide (10 mg) to 2 ml of Affi-Gel 10 in 0.1 M NaHCO₃, pH 8.0, with gentle agitation at 4°C overnight.

2. Block the unreacted coupling sites by the addition of 0.1 ml of 1 M ethanolamine, pH 8.0, for 2 h at room temperature.

3. Pass 1.5 ml of antiserum, which is obtained from the rabbit injected with the peptide BSA conjugate (see *Protocol 9*), over 0.8 ml of a protein A-Sepharose CL-4B column that is equilibrated in PBS.

4. Wash the column with PBS, and elute the bound antibody with 0.1 M citrate buffer, pH 2.5, containing 1 M NaCl, and then dialyse the eluate against PBS at 4°C overnight.

5. Subject the antibody eluted from the protein A column to peptide affinity column chromatography.

6. Mix the antibody in approximately 5 ml of PBS and 1 ml of peptide affinity gel are mixed, and gently agitated at 4°C overnight.

7. Pack this mixture into a column containing 1 M NaCl. Dialyse the solution against PBS for three times every 8 h.

4.2 Analysis of hetero-tetrameric structure of Ins(1,4,5)P_3 receptors

---

Protocol 11. Immunoprecipitation by using type-specific antibodies

Equipment and reagents
- Buffer: 1 mM EDTA, 0.1 mM PMSF, 10 μM leupeptin, 10 μM pepstatin A, 1 mM 2-mercaptoethanol, and 50 mM Tris-HCl, pH 7.4
- ECL Western blotting system (Amersham Corp.)
- Pansobin (Calbiochem-Novabiochem Corp.)

Method

1. Solubilize the membrane proteins (see *Protocol 1*) by addition of 10% Triton X-100 to give final protein and detergent concentrations of 3.0 mg ml^{-1} and 1.0%, respectively.

2. Stir the solution for 30 min at 4°C and centrifuge at 20 000 $\times$ *g* for 60 min at 4°C.

3. Use the supernatant for immunoprecipitation. In non-denaturing conditions, add the solubilized membrane proteins to immunoprecipitation buffer containing 0.15 M NaCl, 0.3% Triton X-100, 0.1% BSA, 5 mM EDTA, 0.1 mM PMSF, 10 μM pepstatin A, and 10 mM sodium phosphate, pH 7.2, and pre-clear with Pansorbin.

4. Incubate the pre-cleared supernatants with 6 μg ml^{-1} antibody against type 1, 2 and 3 Ins(1,4,5)P_3 receptors (1st antibody) for 1 h at 4°C.

5. Add 6 μg ml^{-1} anti-1st antibody IgG (2nd antibody).

6. Collect the immune complexes with Pansorbin.

7. Wash the Pansorbin particles three times with 0.01% bovine serum albumin, 0.5% Triton X-100, 0.15 M NaCl, and 10 mM sodium phosphate, pH 7.2.

8. Mix the Pansorbin pellets with SDS-PAGE sampling buffer containing 4% SDS, 10% 2-mercaptoethanol, 20% glycerol, and 0.125 M Tris-HCl, pH 6.8, and then boil for 3 min.

9. After centrifugation, apply the supernatants to 5% SDS-PAGE in the buffer system of Laemmli (22).

10. After transferring the proteins electrophoretically to nitrocellulose filters, detect the proteins with type-specific antibodies using ECL Western blotting system.

11. In denaturing conditions, boil the solubilized membrane proteins for 5 min in buffer A containing 6 M urea, 20 mM dithiothreitol, 1% SDS, and 1% Triton X-100.

---

12. Dialyse against 0.15 M NaCl, 5 mM EDTA, 0.1 mM PMSF, 10 μM leupeptin, 10 μM pepstatin A, and 10 μM sodium phosphate, pH 7.2.

13. Dilute the concentrations of SDS by 3.3-fold of the sample with 0.15 M NaCl, 5 mM EDTA, 0.1 mM PMSF, 10 μM leupeptin, 10 μM pepstatin A, and 10 μM sodium phosphate, pH 7.2. Subjected these samples to the same immunoprecipitation experiments as the non-denaturing sample.

4.3 Immunohistochemistry

$Ins(1,4,5)P_3$ receptor is ubiquitously distributed through all tissues, although the density differs from cell to cell. It is now known that the three types of receptor are expressed differently in terms of tissue and cell distribution, and in developmental profile. It is important to visualize the localization of each receptor in each tissue. In addition, it was recently found that the distribution of receptors differs even over a single cell. Immunocytochemistry in therefore important to understand the function of $Ins(1,4,5)P_3$ receptors involved in Ca^{2+} transients inside the cell.

There are various methods for the immunohistochemistry to localize the $Ins(1,4,5)P_3$ receptor. Here, I describe the relatively simple method which we use routinely for $Ins(1,4,5)P_3$ receptor immunohistochemistry (*Protocol 12*).

Protocol 12. Immunohistochemistry

Reagents

- Rabbit normal serum (Vector)
- Fixative: 2% paraformaldehyde, 0.025 mol of L-lysine, and 0.01 mol of periodate in phosphate buffer (pH6.2)
- Biotinylated anti-rat IgG (Vector)
- Fluorescein isothiocyanate-conjugated avidin (Vector)

Method

1. Anesthetize mice (adult mice, or pregnant mice) with ether and take out the brain. In case of embryo, the day of conception is determined by the presence of a vaginal plug (embryonic day 0(E0)).

2. Put the E8-E13 embryos in fixative overnight at 4°C. Embryos at stages later than E14, pups, and adult mice are all fixed by cardiac perfusion.

3. Dissect the brain and postfix in the same fixative overnight at 4°C

4. Infiltrate the brain with 20% sucrose in 0.1 mol of phosphate buffer (pH, 7.4).

5. Cut the sample in six-micron-thick section on a cryostat.

6. Mount the sections on gelatin-coated slides and store at −20°C.

191

Protocol 12. *Continued*

7. Incubate the sections with rabbit normal serum (1:30 dilution) in Tris-buffered saline (TBS) for 30 min at room temperature.

8. Incubat the section with the anti-Ins(1,4,5)P3 receptor antibodies (rat IgG) at room temperature for 2 h.

9. Wash the sections with TBS twice.

10. Incubate with biotinylated rabbit anti-rat immunoglobulin antiserum (Vector; 1:200 dilution) for 1 h at room temperature, then rinse twice with TBS and incubate with fluorescein isothiocyanate-conjugated avidin in 0.15 mol of HEPES buffer (1:200 dilution) for 30 min at room temperature.

11. Rinse the section with TBS and mount in buffered glycerol containing 0.1% paraphenylene diamine.

12. Examine with an optic microscope (Nikon) equipped with epi-illumination optics and appropriate filters, or a confocal microscope (Bio-Rad, MRC-500) equipped with epi-illumination optics.

Acknowledgements

I thank Dr Takayuki Michikawa for helping to prepare and Miss Kaori Kawamoto for arranging and typing the manuscript.

References

1. Mikoshiba, K., Huchet, M., and Changeux, J.P. (1979). *Develop. Neurosci.*, **2**, 254.
2. Mikoshiba, K. (1993). *Trends. Pharmacol. Sci.*, **14**, 86.
3. Walaas, S.I., Nairn, A.C., and Greengard, P. (1986). *J. Neurosci.*, **6**, 954.
4. Groswald, D.E. and Kelly, P.T. (1984). *J. Neurochem.*, **42**, 534.
5. Kashiwamata, S., Aono, S., and Semba, R.K. (1980). *Experientia*, **36**, 1143.
6. Supattapone, S., Worley, P.F., Baraban, J.M., and Snyder, S.H. (1988). *J. Biol. Chem.*, **263**, 1530.
7. Maeda, N., Niinobe, M., Nakahira, K., and Mikoshiba, K. (1988). *J. Neurochem.*, **51**, 1724.
8. Maeda, N., Niinobe, M., Inoue, Y., and Mikoshiba, K. (1989). *Develop. Biol.*, **133**, 67.
9. Furuichi, T., Yoshikawa, S., Miyawaki, A., Wada, K., Maeda, N., and Mikoshiba, K. (1989). *Nature*, **342**, 32.
10. Yamamoto-Hino, M., Sugiyama, T., Hikichi, K., Mattei, M.G., Hasegawa, K., Sekine, S., Sakurada, K., Miyawaki, A., Furuichi, T., Hasegawa, M., and Mikoshiba, K. (1994). *Receptors and Channels*, **2**, 9.
11. Maranto, A.R. (1994). *J. Biol. Chem.*, **269**, 1222.
12. Südhof, T.A.C., Newton, C.L., Archer III, B.T., Ushkaryov, Y.A., and Mignery, G.A. (1991). *EMBO J.*, **10**, 3199.

13. Blondel,O., Takeda, J., Janssen, H., Seino, S., and Bell, G.I. (1993). *J. Biol. Chem.*, **268**, 11 356.
14. Furuichi, T. and Mikoshiba, K. (1995). *J. Neurochem.*, **64**, 953.
15. Yamada, N., Makino, Y., Clark, R.A., Pearson, D.W., Mattei, M.-G., Guénet, J.-L., Ohama, E., Fujino, I., Miyawaki, A., Furuichi, T., and Mikoshiba, K. (1994). *Biochem. J.*, **302**, 781.
16. Harnick, D.J., Jayaraman, T., Ma, Y., Mulieri, P., Go, L.O., and Marks, A.R. (1995). *J. Biol. Chem.*, **270**, 2833.
17. Mignery, G.A., Newton, C.L., Archer III, B.T., and Südhof, T.C. (1989). *J. Biol. Chem.*, **265**, 12 679.
18. Fujino, I., Yamada, N., Miyawaki, A., Hasegawa, M., Furuichi, T., and Mikoshiba, K. (1995). *Cell Tissue Res.*, **280**, 201.
19. Maeda, N., Niinobe, M., and Mikoshiba, K. (1990). *EMBO J.*, **9**, 61.
20. Nakade, S., Rhee, S.K., Hamanaka, H., and Mikoshiba, K. (1994). *J. Biol. Chem.*, **269**, 6735.
21. Hirota, J., Michikawa, T., Miyawaki, A., Furuichi, T., Okura, I., and Mikoshiba, K. (1995). *J. Biol. Chem.*, **270**, 19 046.
22. Laemmli, U.K. (1970). *Nature*, **227**, 680.
23. Otsu, H., Yamamoto, A., Maeda, N., Mikoshiba, K., and Tashiro, Y. (1990). *Cell Struct. Funct.*, **15**, 163.
24. Satoh, T., Ross, C.A., Villa, A., Supattapone, S., Pozzan, T., Snyder, S.H., and Meldolesi, J. (1990). *J. Cell Biol.*, **111**, 615.
25. Maeda, N., Kawasaki, T., Nakade, S., Yokota, N., Taguchi, T., Kasai, M., and Mikoshiba, K. (1991). *J. Biol. Chem.*, **266**, 1109.
26. Meyer, T., Wensel, T., and Stryer, L. (1990). *Biochemistry*, **29**, 32.
27. Kiehm, D.J. and Ji, T.H. (1977). *J. Biol.Chem.*, **252**, 8524.
28. Monkawa, T., Miyawaki, A., Sugiyama, T., Yoneshima, H., Yamamoto-Hino, M., Furuichi, T., Saruta, T., Hasegawa, M., and Mikoshiba, K. (1995). *J. Biol. Chem.*, **270**, 14 700.
29. Richardson, C.D., Berkovich, A., Rozenblatt, S., and Bellini, W.J. (1986). *J. Virol.*, **54**, 186.
30. Furuichi, T., Kohda, K., Miyawaki, A., and Mikoshiba, K. (1994). *Current Opinion in Neurobiol.*, **4**, 294.

11

Purification and assay of inositol hexakisphosphate kinase and diphosphoinositol pentakisphosphate kinase

SUSAN M. VOGLMAIER, MICHAEL E. BEMBENEK,
ADAM I. KAPLIN, GYÖRGY DORMÁN, JOHN D. OLSZEWSKI,
GLENN D. PRESTWICH, and SOLOMON H. SNYDER

1. Introduction

The identification of diphosphoinositol pentakisphosphate (PP-InsP_5) and bisdiphosphoinositol tetrakisphosphate ([PP]$_2$-InsP_4), inositol pyrophosphates with seven and eight phosphates associated with the inositol ring (1–3), indicates that inositol hexakisphosphate (InsP_6) is not the endpoint of *myo*-inositol metabolism with a slow turnover rate as was once thought. An active substrate cycle has been revealed by phosphatase inhibition in which up to 50% of the 15–60 μM pool of InsP_6 in cells is converted to pyrophosphate derivatives each hour (4). An InsP_6 kinase has been purified and can be dissociated from a PP-InsP_5 kinase activity. The pure InsP_6 kinase displays high affinity and selectivity for InsP_6 as substrate and, in the reverse direction, transfers a phosphate from PP-InsP_5 to ADP to form ATP (5). This ATP synthase activity demonstrates the high phosphoryl group transfer potential of PP-InsP_5, which may represent a physiological role for PP-InsP_5, either by generating ATP or by transferring its energy directly. These molecules represent a new class of inositol phosphates containing high-energy pyrophosphate bonds.

Investigation of the complex metabolism of these higher inositol phosphates may contribute to a greater understanding of their roles in cell function. In plants, InsP_6 acts as an antioxidant (6) and phosphate store (7). In cells, InsP_6 may act as a siderophore (8, 9) and thus block iron-mediated oxidative damage (10). InsP_6 also interacts with several proteins which regulate endocytosis (11–15), synaptic vesicle trafficking (16–18) and receptor desensitization (19). The high rate of turnover of PP-InsP_5 (4) and its regulation by intracellular calcium (20) suggest that interconversion of $InsP_6$ and $PP\text{-}InsP_5$ and/or $[PP]_2\text{-}InsP_4$ would be an excellent candidate energy source and/or regulatory switch for processes involving these proteins.

2. Purification of $InsP_6$ kinase

2.1 Tissue source

In a screen of various tissues, highest concentrations of enzyme activity are evident in the brain and testis. The thymus possesses about 60% of brain activity while enzyme activity in heart, liver and kidney is 5–25% of brain values. Because of the abundance of phosphatase activity in most tissues, these estimates may not reflect the absolute levels of the $InsP_6$ kinase. In brain extracts, virtually all $InsP_6$ kinase activity is recovered in the soluble supernatant fraction after $100\,000 \times g$ centrifugation.

2.2 Properties and stability

SDS-gel electrophoresis of the purified enzyme reveals an apparent molecular weight of 54 kDa. The elution profile of the enzyme by size exclusion chromatography is consistent with a molecular weight of about 60 kDa, which implies that $InsP_6$ kinase is a monomer. The enzyme loses 50% of its activity when stored for 7 days at 4°C. Optimal stability over at least 6–8 weeks is evident when the enzyme is stored with 20% glycerol at −70°C. Enzyme activity displays a broad pH optimum with maximal activity at pH 6.8. A variety of buffers can be used, HEPES being optimal.

K_m values (assayed by *Protocol 2*) of the purified $InsP_6$ kinase for $InsP_6$ and ATP are 0.7 µM and 1.35 mM respectively. At concentrations of $InsP_6$ > 50 µM and ATP > 10 mM, substrate inhibition is observed. For the reverse reaction, the K_m values (assayed by *Protocol 3*) for $PP\text{-}InsP_5$ and ADP are 1.97 µM and 1.57 mM, respectively. The V_{max} values for the forward and reverse reactions are 1.41 and 2.64 $\mu mol\ min^{-1}\ mg^{-1}$ respectively. Kinetic analysis reveals both the forward and reverse reactions to be random bireactant systems.

2.3 Purification

Enzyme activity adsorbs to a heparin column and is eluted with KCl providing a 37-fold purification, while purification with the anion exchanger column Mono Q provides a further 4-fold enrichment of enzyme activity. The next greatest purification is obtained by adsorbing the enzyme to an $InsP_6$ affinity column, affording a 13-fold increase in specific activity. Adsorption to a second heparin column with a gradient elution provides another 5- to 6-fold purification, and a gel filtration column results in a further 3- to 4-fold purification. The final preparation is enriched 40 000-fold in enzyme activity and provides a 2.2% yield.

The fluoride-sensitive phosphatase activity flows through the first heparin column. PP-InsP_5 kinase activity copurifies with InsP_6 kinase activity on the first heparin agarose column (step 4). Ins(1,4,5)P_3 kinase activity is separated by the second heparin-agarose column (step 7). The purification procedure may be scaled down.

Protocol 1. Purification of InsP_6 and PP-InsP_5 kinase activities

Equipment and reagents

- Rat forebrain
- Buffer A: 20 mM HEPES (pH 6.8) (Research Organics), 4 mM DTT, 2 mM EGTA, 0.75 mM EDTA, 5 mM NaF, 1.5 mM Na$_3$VO$_4$, 0.5 mg litre^{-1} okadaic acid, 4 mg litre^{-1} chymostatin, 4 mg litre^{-1} pepstatin, 4 mg litre^{-1} antipain, 8 mg litre^{-1} leupeptin, 8 mg litre^{-1} aprotinin, and 200 mg litre^{-1} phenylmethyl-sulfonylfluoride
- Buffer B: 20 mM HEPES (pH 6.8), 1 mM DTT, 1 mM EGTA, 2 mg litre^{-1} chymostatin, 2 mg litre^{-1} pepstatin, 2 mg litre^{-1} antipain, 4 mg litre^{-1} leupeptin, and 0.1% CHAPS

- Heparin-agarose, Type 1 (Sigma H6508)
- Mono Q HR 10/10 FPLC column (Pharamacia)
- IP$_6$-Affiprep resin (prepared as in ref. 21)
- Centriprep-30 concentrator (Amicon)
- Superdex 75 gel filtration column (Pharamacia)
- InsP_6 (Calbiochem)

Method

1. Homogenize the forebrains (70 g wet weight) from 70 adult male (175–300 g) Sprague-Dawley rats in 150 ml ice-cold Buffer A (or 1 rat forebrain in 2.5 volumes).

2. Centrifuge at 100 000 × *g* for 90 min.

3. Agitate the supernatant with 70 ml heparin-agarose (1 ml resin per 1 gram wet weight tissue) in the presence of 1 mM MgCl$_2$ and 0.1% CHAPS for 60 min.

4. Wash the resin with 450 ml (6–7 column volumes) 0.25 M KCl in Buffer A with 1 mM MgCl$_2$ minus okadaic acid, then elute by agitating for 60 min with 75 ml (approx. 1 column volume) 0.5 M KCl at pH 7.4 with 1 mM EDTA in Buffer A minus all phosphatase inhibitors. Pour heparin agarose into a column, elute, and then elute with an additional 50 ml (approx. 1 column volume) of the same buffer.

5. Dilute the combined heparin column eluate to 1 litre in Buffer B and apply to a Mono Q HR 10/10 FPLC column at 4 ml min^{-1}, wash with 50 ml 0.05 M KCl, and elute at 0.5 ml min^{-1} with a 60 ml linear KCl gradient from 0.05 to 0.45 M KCl in Buffer B plus 0.6% CHAPS.

6. Assay the InsP_6 kinase activity (*Protocol 2*) and dilute the pooled peak of activity to 0.05 M KCl, load onto a 1 × 8 cm InsP_6-Affiprep column at 0.5 ml min^{-1} and pass the flow-through over the column again. Wash the column with 50 ml 0.15 M KCl and elute with 16 ml 25 μM InsP_6 in Buffer B plus 0.6% CHAPS.

Protocol 1. *Continued*

7. Dilute the pooled peak of activity to 2 μM InsP_6 and apply to a 0.5 × 4 cm heparin agarose column at 0.5 ml min^{-1}, wash with 20 ml 0.3 M KCl and elute at 0.2 ml min^{-1} with a 20 ml linear gradient from 0.3 M to 0.6 M KCl in Buffer B plus 0.6% CHAPS.

8. Concentrate the peak fractions from the heparin column in a Centriprep-30 concentrator (Amicon) to 100 μl and apply to a Superdex 75 gel filtration column run at 0.2 ml min^{-1} in Buffer B plus 0.6% CHAPS.

[a]If tissue supernatant is to be used directly in the InsP_6 kinase assay, Na$_3$VO$_4$ may be omitted, diluted, or dialysed away as it inhibits InsP_6 kinase activity.

3. Polyethyleneimine-cellulose thin layer chromatography

3.1 Assay of InsP_6 and PP-InsP_5 kinase activities

We have adapted methods that separate InsP_6, InsP_5, InsP_4, InsP_3, ATP, and P$_i$ using polyethyleneimine-cellulose thin layer chromatography (PEI-TLC) (22, 23) to efficiently separate InsP_6, PP-InsP_5, [PP]$_2$-InsP_4 and ATP, with no overlap between these phosphates (*Figure 1*). InsP_6 kinase and PP-InsP_5 activity is monitored by measuring [^{3}H]PP-InsP_5 formed from [^{3}H]InsP_6 and [^{3}H][PP]$_2$-InsP_4 formed from [^{3}H]PP-InsP_5.

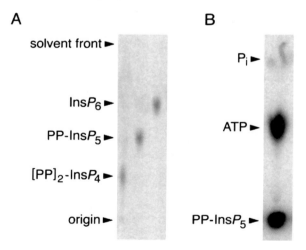

Figure 1. (A) Separation of ^{3}H-labelled inositol hexakisphosphate (InsP_6), diphospho-inositol pentakisphosphate (PP-InsP_5) and bisdiphosphoinositol tetrakisphosphate ([PP]$_2$-InsP_4) by polyethyleneimine-cellulose thin layer chromatography (PEI-TLC) (*Protocol 2*). (B) Separation of [^{32}P]PP-InsP_5 and [γ-^{32}P]ATP by PEI-TLC (*Protocol 3*).

Protocol 2. Assay of InsP_6 and PP-InsP_5 kinase activity

Equipment and reagents

- Reaction buffer: 20 mM HEPES (pH 6.8), 1 mM DTT, 6 mM MgCl$_2$, 5 mM Na$_2$ATP, 5 mM NaF, 10 mM phosphocreatine, 40 U ml^{-1} creatine phosphokinase (Sigma), and 0–10 μM InsP_6 (Calbiochem) or PP-InsP_5 (NEN/DuPont)
- [^{3}H]InsP_6, [^{3}H]PP-InsP_5 and [^{3}H][PP]$_2$-InsP_4 (NEN/DuPont)
- TLC solvent: 1.1 M KH$_2$PO$_4$, 0.8 M K$_2$HPO$_4$ (Aldrich), and 2.3 M HCl[a]
- Polyethyleneimine-cellulose TLC (PEI-TLC) plates (EM Separations)
- Hair dryer

Method

1. Incubate 1–4 μl enzyme preparation from *Protocol 1* or tissue supernatant in 10 μl final volume of reaction buffer with 40–80 nM of either [^{3}H]InsP_6 (InsP_6 kinase assay) or [^{3}H]PP-InsP_5 (PP-InsP_5 kinase assay) at 37°C for 10–30 min.

2. Terminate the reaction either by adding 1 μl 1 M HCl or by immersing in an ice water bath.

3. Spot each sample in 2.5 μl aliquots onto a PEI-TLC plate at 1.25 cm intervals and dry under warm air.

4. Develop the TLC plates in 1.1 M KH$_2$PO$_4$, 0.8 M K$_2$HPO$_4$, and 2.3 M HCl (approximately 2.0–2.5 h).

5. Dry the plates under warm air.

6. To measure the R_f values[b], cut a lane containing reaction mixture, enzyme preparation, and standards of [^{3}H]InsP_6, [^{3}H]PP-InsP_5, and/or [^{3}H][PP]$_2$-InsP_4 into 1 cm sections and count with 15 ml DuPont Formula 963 scintillation cocktail.

7. Cut out the appropriate region in each lane and count as above, or shake the TLC segments with 3 ml of concentrated HCl, then count with 5 ml of H$_2$O and 10 ml scintillation cocktail.[c]

[a]Heat potassium phosphate to dissolve, then add HCl.
[b][^{3}H]InsP_6, [^{3}H]PP-InsP_5, and [^{3}H][PP]$_2$-InsP_4 migrate with R_f values of approximately 0.75, 0.45, and 0.3, respectively, but the migration of standards should be measured with each experiment. Glycerol, phospholipids, and high protein concentrations will change R_f values.
[c]Approximately 25% of the added ^{3}H applied to the plates is recovered by the first method and up to 80% by the second method.

3.2 Assay of PP-InsP_5 and [PP]$_2$-InsP_4 phosphatase activity

PP-InsP_5 and [PP]$_2$-InsP_4 phosphatase activity may be assayed in tissue supernatants using *Protocol 2*, omitting the ATP, phosphocreatine, creatine phosphokinase, and NaF. Alternately, if [^{32}P]-labelled substrates are available

(NEN/DuPont), phosphatase activity can be assayed by measuring [^{32}P] release using ion-exchange chromatography [4].

3.3 Assay of ATP synthase activity by InsP_6 kinase

When the kinase is incubated in the presence of [^{32}P]PP-InsP_5 and unlabelled ADP, [^{32}P]ATP is formed. The reaction is highly selective for ADP as GDP, CDP, UDP, AMP, creatine, glucose, and 3-phosphoglycerate are all ineffective (5). The partially purified [PP]$_2$-InsP_4 kinase also specifically forms [^{32}P]ATP from [^{32}P][PP]$_2$-InsP_4 and unlabelled ADP (unpublished data).

Protocol 3. Assay of ATP synthase activity

Equipment and reagents

- Reaction buffer: 20 mM HEPES (pH 6.8), 1 mM DTT, 5 mM Na$_2$ADP, 5 mM MgCl$_2$, 0–10 μM PP-InsP_5 or [PP]$_2$-InsP_4 (NEN/DuPont)
- [^{32}P]PP-InsP_5 for [^{32}P][PP]$_2$-InsP_4, and [γ-^{32}P]ATP (NEN/DuPont)
- TLC solvent 1.0 M KH$_2$PO$_4$
- Polyethyleneimine-cellulose TLC (PEI-TLC) plates
- Hair dryer

Method

1. Incubate 1–4 μl enzyme preparation from *Protocol 1* or tissue supernatant in 10 μl final volume of reaction buffer with 50 nM [^{32}P]PP-InsP_5 or [^{32}P][PP]$_2$-InsP_4 at 37°C for 10–30 min.

2. Terminate the reaction and spot onto a PEI-TLC plate as in *Protocol 2*.

3. Develop the TLC plates in 1.0 M KH$_2$PO$_4$ (approximately 1.0–1.5 h).

4. Dry the plates under warm air.

5. To measure the R_f values, cut a lane containing reaction mixture, enzyme preparation and standards of [^{32}P]PP-InsP_5, [^{32}P][PP]$_2$-InsP_4, and [γ-^{32}P]ATP into 1 cm sections and count with 15 ml DuPont Formula 963 scintillation cocktail. Alternatively, the TLC plate may be exposed to film or the radioactivity may be localized with a hand held β counter.

6. Cut out the appropriate region in each lane and count as in *Protocol 2*.

[a][^{32}P]PP-InsP_5, [^{32}P][PP]$_2$-InsP_4, and [γ-^{32}P]ATP migrate with R_f values of approximately 0.05, 0.05, and 0.55, respectively, but the migration of standards should be measured with each experiment.

References

1. Mayr, G.W., Radenberg, T., Thiel, U., Vogel, G., and Stephens, L.R. (1992). *Carbohydrate Research*, **234**, 247.
2. Menniti, F.S., Miller, R.N., Putney, J.W., Jr., and Shears, S.B. (1993). *J. Biol. Chem.*, **268**, 3850.

3. Stephens, L., Radenberg, T., Thiel, U., Vogel, G., Khoo, K.H., Dell, A., Jackson, T.R., Hawkins, P.T., and Mayr, G.W. (1993). *J. Biol. Chem.*, **268**, 4009.
4. Shears, S.B., Craxton, A.N., and Bembenek, M.E. (1995). *J. Biol. Chem.*, **270**, 10 489.
5. Voglmaier, S.M., Bembenek, M.E., Kaplin, A.I, Dorman, G., Olszewski, J.D., Prestwich, G.D., and Snyder, S.H. (1996). *Proc. Natl. Acad. Sci. USA*, **93**, in press.
6. Graf, E., Empson, K.L., and Eaton, J.W. (1987). *J. Biol. Chem.*, **262**, 11 647.
7. Morton, R.K. and Raison, J.K. (1963). *Nature*, **200**, 429.
8. Sasakawa, N., Sharif, M., and Hanley, M.R. (1995). *Biochem. Pharmacol.*, **50**, 137.
9. Poyner, D.R., Cooke, F., Hanley, M.R., Reynolds, D.J., and Hawkins, P.T. (1993). *J. Biol. Chem.*, **268**, 1032.
10. Hawkins, P.T., Poyner, D.R., Jackson, T.R., Letcher, A.J., Lander, D.A., and Irvine, R.F. (1993). *Biochem. J.*, **294**, 929.
11. Voglmaier, S.M., Keen, J.H., Murphy, J., Ferris, C.D., Prestwich, G., Snyder, S.H., and Theibert, A.B. (1992). *Biochem. Biophys. Res. Commun.*, **187**, 158.
12. Timerman, A.P., Mayrleitner, M.M., Lukas, T.J., Chadwick, C.C., Saito, A., Watterson, D.M., Schindler, H., and Fleischer, S. (1992). *Proc. Natl. Acad. Sci. USA*, **89**, 8976.
13. Norris, F.A., Ungewickell, E., and Majerus, P.W. (1995). *J. Biol. Chem.*, **270**, 214.
14. Ye, W., Li, N., Bembenek, M.E., Shears, S.B., and Lafer, E.M. (1995). *J. Biol. Chem.*, **270**, 1564.
15. Fleischer, B., Xie, J., Mayrleitner, M., Shears, S.B., Palmer, D.J., and Fleischer, S. (1994). *J. Biol. Chem.*, **269**, 17 826.
16. Llinas, R., Sugimori, M., Lang, E.J., Moeir, M., Fukuda, M., Niinobe, M., and Mikoshiba, K. (1994). *Proc. Natl. Acad. Sci. USA*, **91**, 12 990.
17. Niinobe, A.U., Yamaguchi, Y., Fukuda, M., and Mikoshiba, K. (1994). *Biochem. Biophys. Res. Commun.*, **205**, 1036.
18. Schiavo, G., Gmachi, M.J.S., Stenbeck, G., Sollner, T.H., and Rothman, J.E. (1995). *Nature*, **378**, 733.
19. Palczewski, K., Pulvermuller, A., Buczylko, J., Gutmann, C., and Hofmann, K. P. (1991). *Federation of European Biological Societies*, **295**, 195.
20. Glennon, M. C. and Shears, S. B. (1993). *Biochem. J.* **293**, 583.
21. Marecek, J.F. and Prestwich, G.D. (1991). *Tetrahedron Lett*, **32**, 1863.
22. Spencer, C.E.L., Stephens, L.R., and Irvine, R.F. (1990). In *Methods in inositide research* (ed. R.F. Irvine), p. 39. Raven Press, New York.
23. Ryu, S.H., Lee, S.Y., Lee, K.Y., and Rhee, S.G. (1987). *FASEB J.*, **1**, 388.

12

Phosphoinositide 3-kinase

SEBASTIAN PONS, DEBORAH J. BURKS, and
MORRIS F. WHITE

1. Introduction

Plants, yeast and animal cells contain phosphoinositide 3-kinase activities (PI3K), each of which is capable of transferring the terminal phosphate of ATP to the D-3 position of the inositol head groups of one or more of the substrates PtdIns, PtdIns4P, and PtdIns(4,5)P_2 (see Chapter 1 and references therein). PI3K is a downstream effector of several receptor tyrosine kinases and plays a central role in a broad range of biological effects (1). PI3K activation can potentially generate three products: PtdIns3P, PtdIns(3,4)P_2, and PtdIns(3,4,5)P_3 (2), with the last of these regarded as the most important signal-generated product (*Figure 1*). PI3K is a heteromeric enzyme consisting of a catalytic subunit and a regulatory subunit. The first examples of these subunits were described as p110α and p110β (catalytic) and p85α and p85β (regulatory). The p85 regulatory subunits have no known catalytic activity but they possess a number of regions believed to be important in protein–protein interactions, i.e. two Src-homology 2 (SH2) domains, an SH3 domain and a region of homology to the breakpoint cluster region gene (*bcr*) which may interact with and regulate the rho family of small-molecular weight GTP-binding proteins (1) (*Figure 2A*). The SH2 domains of p85 bind phosphotyrosine residues in the motif YXXM or YMXM (3). Many growth factor receptors contain YMXM amino-acid motifs and bind PI3K directly, but the insulin receptor and various cytokine receptors engage PI3K via phosphotyrosine motifs of insulin receptor substrate (IRS) molecules (4, 5). A region between the two SH2 domains of p85 binds to the N-terminus of p110, partially activating the catalytic subunit. Although the p85α and p85β are highly homologous, evidence suggests that functional differences may exist between these isoforms.

Recently, new members of the PI3K regulatory family have been identified. By screening expression libraries with tyrosine phosphorylated insulin receptor substrate (IRS-1), we isolated a novel 55 kDa PI3K regulatory subunit (p55^{PIK}) (6). Like known p85 isoforms, p55^{PIK} contains two SH2 domains and a p110 interaction site (*Figure 2A*). However, p55^{PIK} lacks an N-terminal

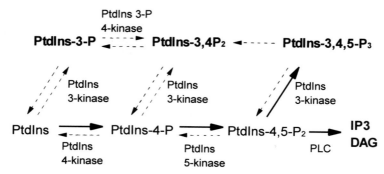

Figure 1. Metabolism of phosphoinositides. The lipids generated by the 3′-phosphoinositide pathway are in bold to distinguish them from the phosphoinositides involved in the canonical inositol lipid turnover pathway. The solid arrows indicate pathways known to occur *in vivo*, the dashed arrows indicate reactions that may only occur *in vitro*. The enzymes that catalyse the different reactions are indicated. PLC, phospholipase C; DAG, diacylglycerol.

SH3 domain and a BCR-like region; instead, it has a unique 30-amino acid N-terminus with a potential tyrosine phosphorylation site and a proline-rich motif. Using a similar screening strategy with phosphorylated IRS-1, another PI3K regulatory molecule has been identified; p55α is proposed as an alternative splicing variant of the p85α gene (*Figure 2A*) (7).

Since the original cloning of the PI3K catalytic subunit p110α, several other p110 isoforms have been cloned, including p110β and recently p110γ (8, 9). While the p110α and p110β are relatively similar and contain common structural motifs, p110γ does not bind to regulatory subunits (p85 or p55) and differs by the addition of a pleckstrin homology (PH) domain at its N-terminus (*Figure 2B*) (10). At least some PH domains have been documented to bind inositol lipids and/or βγ subunits of heterotrimeric G-proteins, so the PH domain of p110γ may provide a mechanism for mediating subcellular translocation and activation of PI3K.

PI3K has been implicated in the regulation of various cellular activities, including proliferation (11), differentiation (12), membrane ruffling (13), regulated secretion (14), and the prevention of apoptosis (15). In addition, PI3K activation is required for insulin-stimulated glucose uptake and insulin-mediated p70^{S6K} activity (16, 17). Insulin and IGF-I stimulate the PI3K activity through tyrosine phosphorylation of IRS proteins, although there is evidence suggesting that these two receptors may also directly associate with p85 (4). Recent reports suggest that PI3K may be regulated by p21ras; overexpression of activated p21ras in PC12 cells stimulates PI3K activity, resulting in accumulation of 3′-phosphorylated inositol lipids in the cells (18). Furthermore, GTP-bound p21ras binds and directly activates PI3K *in vitro* (18). However, in other systems PI3K-mediated events such as

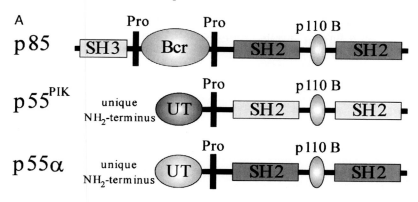

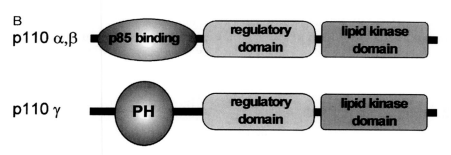

Figure 2. Structure of the PI3K regulatory and catalytic subunits. (A) Comparison of the various PI3K regulatory subunits, depicting the conserved structural domains. p85 and p55α are splicing variants of the same gene, whereas p55PIK is the unique product of a different gene. (B) Comparison of p110α, β and the recently cloned γ. The three molecules are very similar at the C-terminus, containing a lipid kinase domain and a regulatory region. However, they differ at the N-terminus; p110 α and β both have a p85-p55 binding domain but p110γ instead contains a pleckstrin homology (PH) domain.

activation of $p70^{s6K}$ and glucose transport do not appear to require p21ras activation (19, 20).

Much of the progress in understanding PI3K as an upstream element in signalling cascades has been made possible by the availablity of inhibitors of the enzyme, in particular the fungal metabolite wortmannin (21). At concentrations up to 100 nM, wortmannin irreversibly inhibits both the lipid and serine kinase activities of PI3K, through covalent interaction with the p110 catalytic site (14, 22). However, recent evidence underscores the problems associated with interpreting results that are based solely on the use of pharmacological inhibitors such as wortmannin; at concentrations that are frequently used experimentally for studying PI3K, wortmannin inhibits at least two other enzymes, namely PtdIns 4-kinase and PLA2 (23, 24). Moreover, at higher concentrations (> 1 μM) wortmannin inhibits other enzymes and signalling pathways. Thus, experimental results interpreted to be the effect of wortmannin on PI3K may in fact represent the cumulative effect of this reagent on several cellular responses and biochemical events within the stimulated cell.

Although recent studies have revealed much regarding the regulation of PI3K by upstream signals, the mechanisms by which PI3K products actually generate a signal is not fully understood, especially as they are present in very small quantities within the cell. One critical aspect of studying the role of PI3K in cellular signalling is the measurement of its activity. The protocols which have been established to measure the activity of PI3K can be classified in terms of two basic groups: (a) *in vivo* assays where live cells are labelled with either [^{3}H]inositol or ^{32}P-P$_i$ and the appearance of radiolabelled PI3K products is measured; and (b) *in vitro* assays in which cell lysates are prepared and the PI3K activity present in immunocomplexes is evaluated by measuring incorporation of ^{32}P into an exogenous substrate. In this chapter, we present an overview of the methodology for measuring PI3K activity and compare the advantages and disadvantages of these two approaches.

2. Assays for PI3K activity

2.1 *In vivo* measurements

It is quite time-consuming to measure changes in the rate of synthesis of PtdIns3P, PtdIns(3,4)P_2 and PtdIns(3,4,5)P_3 in intact cells in response to a particular stimulus. However, such an *in vivo* assay can provide detailed information about the regulation of the PI3K activity. Since the mass levels of these inositol lipids are quite small, analysis of changes in their levels generally requires a radiolabelling strategy. [^{3}H]inositol labelling protocols are described in Chapters 1 and 2. Here, we focus on the strategy of pre-labelling the cellular ATP pool with [^{32}P], which has the advantage of being accomplished on a shorter time scale (up to 90 min); this is performed in serum-free conditions. Following stimulation with the factor of interest, cells are immediately solubilized with organic solvents and products are identified by HPLC. For a better resolution of the products of interest, we strongly recommend a pre-purification on a TLC plate prior to the final separation by HPLC.

The length of serum-deprivation, labelling, and stimulation will depend on the type of cell line and the particular stimulus, and may require some modifications in each case. The method described below (*Protocol 1*) has been used for 3T3L1 cells (17), but can be easily adapted for other types of cells.

Protocol 1. *In vivo* assay of phosphatidylinositides

Equipment and reagents

- Whatman 5 μm Partisphere SAX column (12.5 cm cartridge, Whatman 4611–0505) or similar
- Guard Column (integral guard column system, AX guard cartridge system, Whatman 4651–0005)
- Radiomatic Flow-one-beta on-line radioactive detector

- Buffer A: 30 mM HEPES pH 7.4 containing 110 mM NaCl, 10 mM KCl, 1 mM MgCl, 1.35 mM CaCl, and 10 mM glucose
- Buffer B: $(NH_4)_2HPO_4$, 1 M, pH 3.8 with phosphoric acid

A. *Cell labelling*

1. Starve the cells for 4 h in serum-free media. Wash in buffer A. Resuspend at a density of 5×10^7 cells/ml in the same buffer containing 2 mg ml^{-1} BSA but without Ca^{+2} (in our hands this avoids cell aggregation and improves the degree of stimulation in the 3T3L1 cell line).

2. Add 0.5 mCi of ^{32}P-P$_i$ per ml (3000 Ci mmol^{-1}) to the resuspended cells and incubate at 37 °C for 90 min.

3. Remove the excess extracellular radioisotope by washing 2 times with 10 ml of buffer A.

4. Resuspend cells using Ca^{+2}-free buffer A and divide into aliquots of 1 ml containing 2×10^7 cells/ml. Incubate at 37°C and stimulate for the appropriate times with the ligand of interest.

B. *Lipid extraction*

1. Transfer the stimulated cells to a tube containing 3 ml of CHCl$_3$/MeOH (1:2) plus 1 mg ml^{-1} of butylated hydroxytoluene and 10 μg ml^{-1} of a 1:1:1 mixture of PtdIns/PtdIns4P/ PtdIns(4,5)P_2.

2. Add 2.1 ml of CHCl$_3$ and 2.1 ml of 2.5 M HCl to the tubes and then mix well and centrifuge.

3. Collect the lower phase and wash the upper phase twice with 1 ml of CHCl$_3$. The three upper phases should be pooled and dried under vacuum.

C. *TLC pre-purification*

1. Resuspend dry samples in 100 μl of CHCl$_3$.

2. Spot the lipids on an silica plate (20 × 20 cm 0.2 mm thickness DC-Alufolien Kieselgel 60, supplied by Merck), immersed in 1.2% Potasium Oxalate for 30 min). Separate using chloroform/acetone/methanol/acetic acid/water (80:30:26:24:14).

3. Expose the plate to film for autoradiography.

Protocol 1. *Continue*

4. To determine migration positions of the products, compare with migration of standards of [^{32}P]-PtdIns3P, [^{32}P]-PtdIns(3,4)P_2, [^{32}P]-PtdIns(3,4,5)P_3. These can be made from the appropriate precursors (supplied by Sigma) using immunoprecipitated 3-kinase (see *Protocol 2*).

5. Reveal the position of the unlabelled PtdIns/PtdIns4P/ PtdIns(4,5)P_2 as an assay standard by vaporizing the plate with I_2 as follows: in a fume hood, spray the plate with I_2. Allow to dry. Lipids appear in 10 min.

D. *Deacylation of phosphoinositides*

1. Using the autoradiograph as a template, cut PtdIns3P, PtdIns (3,4)P_2, and PtdIns (3,4,5)P_3 from the plate and place in a 20 ml glass scintillation vial.

2. Add to the vial 1.5 ml of methylamine reagent (57.7 ml 25% methylamine in water, 61.6 ml MeOH, 15.4 ml *n*-BuOH) as well as 1 μl cold PtdIns4P as carrier (10 mg ml^{-1} stock in CHCl$_3$).

3. Incubate at 53 °C for 1 h, then dry by vacuum centrifugation.

4. Add 1ml of water and dry again by vacuum centrifugation.

5. To the dry residue, add 0.6 ml (lower phase) of water and 0.7 ml of *n*-BuOH/light petroleum ether/ethyl formate (20:4:1) (upper phase mix). After vortexing and centrifuging, the mix forms two phases: an upper containing transacylated methylamine and nondeacylated lipids, and the lower phase containing glycerophosphoesters.

6. Discard upper phase and wash the lower phase with 0.7 ml of the upper phase mix.

7. Again discard the upper phase and dry the lower phase by vacuum centrifugation.

8. Resuspend samples in 250 ul of 0.01 M (NH$_4$)$_2$ HPO$_4$, pH 3.8 with phosphoric acid, filter with 0.45 um PVDF filter to eliminate residual particulate silica from the TLC.

9. Add 1 μl of internal standards of [^{3}H]GroPIns4P and [^{3}H]GroPIns(4,5)P_2. These are prepared by adding 0.1 mg + 1 mCi of either [^{3}H]PtdIns4P or [^{3}H]PtdIns(4,5)P_2 to a scintillation vial, and are dried under argon and deacylated as described above.

E. *HPLC separation of the deacylated phosphoinositides*[a]

1. Run samples at room temperature using a Whatman 5 μm Partisphere SAX or similar, plus a Guard Column. Assay in two channel mode (^{32}P-^{3}H) using flow scint IV scintillant with ^{3}H window set at 0–20 and the ^{32}P window set at 21–500. The HPLC flow must be 1 ml min^{-1} and

the scintillant flow 4 ml min^{-1}. Program the HPLC machine as follows: 0 min, 0% B; 60 min, 25% B; 65 min, 100% B, 70 min, 100% B, 75 min, 0% B.

[a] In this final step, the deacylated inositol lipids are resolved using HPLC chromatography. (Chapters 1 and 2 provide more detailed descriptions of the precautions that should be taken during HPLC to improve resolution and reproducibility).

Using the technique described in *Protocol 1*, separation is normally very clean because only phosphotidylinositides should be present in the final extract. This protocol can be used with total phospholipid extracts; however, several other products will then elute prior to PtdIns*P*. To illustrate the results of TLC pre-purification, a representative experiment performed with insulin-stimulated 3T3-L1 cells is shown in *Figure 3*.

2.2 *In vitro* assays

Activity of PI3K can be measured *in vitro* by monitoring the incorporation of ^{32}P into PtdIns. As explained above, the regulatory subunits of PI3K (p85 and p55) associate with many phosphotyrosine-containing signalling molecules. However, both the regulatory and the catalytic subunits of PI3K can also bind to other molecules through phosphotyrosine-independent mechanisms. For example, p85 subunits associate with p59^{fyn} (25) and p110 has been shown to bind p21^{Ras} (18). Activation of the PI3K has been observed to occur due to binding of the regulatory subunits through both phosphotyrosine-dependent and independent mechanisms (18, 26). In contrast, the binding to the catalytic subunit is always independent of phosphotyrosine. In this section, we discuss measuring PI3K activity from two different cellular perspectives: (a) association of PI3K with a certain molecule, and (b) changes in the kinetics of the enzyme which result from agonist-stimulation.

In the first experimental scenario (a), PI3K activity associated with another molecule is measured following co-immunoprecipitation, to determine whether a particular stimulus induces association of PI3K with the molecule of interest. Immunoprecipitations from lysates of unstimulated and stimulated cells are performed using an antibody against the molecule of interest, followed by measurement of PI3K activity in the immunocomplexes. In most cases, use of an antibody against a molecule predicted to associate with PI3K is not a technical concern, since antibody binding to an associated molecule rarely affects activity of the PI3K enzyme. A caveat with this approach is the inability to distinguish between a change in protein–protein interaction and a change in the kinetics of the enzyme resulting from association.

In the second experimental scenario (b), where the interest is in determining whether a stimulus alters the kinetic parameters of the PI3K enzyme, consideration must be given to assay conditions that will accurately deter-

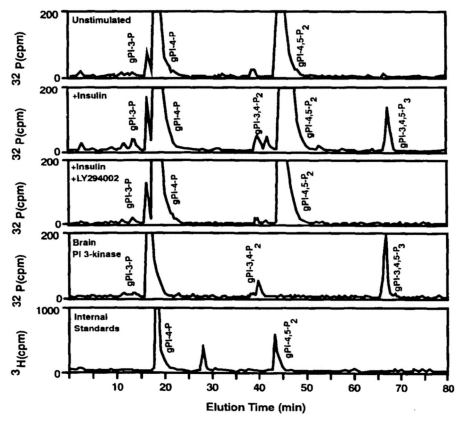

Figure 3. Typical HPLC elution patterns of the different PI3K products generated by *in vivo* labelling. In this experiment, the effect of insulin and the PI3K inhibitor LY294002 on PI3K activity was measured in 3T3L1 cells. The upper three panels represent the phosphoinositides found in unstimulated, insulin and insulin plus LY294002 treated cells, respectively. The lower two panels represent the elution pattern of the phosphoinositides generated with a purified brain PI3K and of the commercial tritiated phospholipids used as standards. Reproduced with permission from ref. 17.

mine the kinetic objectives of the assay. One consideration is whether to immunoprecipitate using antibodies directed against the regulatory or catalytic subunit of PI3K. Since the regulatory and the catalytic subunits form stable complexes that do not appear to be regulated, it should be possible, in theory, to immunoprecipitate with any antibody against either subunit. However, in practice, very few antibodies are useful for measuring changes in the kinetics of PI3K. Antibodies to the catalytic subunits often themselves modify the kinetic parameters, and those raised against the regulatory subunits must be based on regions of the molecule that do not participate in interactions with the catalytic subunit. In our laboratory, we have tested several antibodies to p85 or p55 and have found that those generated against either a peptide comprising amino acids 144–160 of p85α or the first 30 amino acids of p55^{PIK}, do not effect the activity of the PI3K. In contrast, antibodies made against the SH2 domains or the p110 binding site activate the catalytic subunit.

In either experimental situation, whether immunoprecipitating with an antibody against an associated molecule or one against the PI3K itself, the protocol presented below has been used extensively in our laboratory to measure the PI3K activity associated with IRS and to monitor changes in the enzyme activity resulting from this association (26).

Protocol 2. Measurement of PI3K activity associated with insulin receptor substrate

Method

1. Starve cell cultures for at least 4 h in serum-free media, stimulate with an appropriate agonist for various times and immediately extract in lysis buffer (1 ml per 10 cm dish of 20 mM Tris pH 7.4, 137 mM NaCl, 1 mM MgCl$_2$, 1 mM CaCl$_2$, 10% glycerol, 1% NP40 and freshly added 1 μg ml^{-1} leupeptin, 1 μg ml^{-1} aprotinin, 25 mM PMSF and 100 μM orthovanadate).

2. Remove cells from plate by scraping and transfer to a 1.5 ml microfuge tube.

3. Centrifuge at 13 000 r.p.m. for 15 min to remove the NP40 insoluble material.

4. Incubate supernatants for 4 h. with the appropriate antibodies (see text)

5. Add 30 μl of Protein A Sepharose beads to the tube. Following incubation for an additional hour on a rotating platform, tubes are centrifuged and sepharose beads washed as follows.

    • 3 washes with 1 ml PBS pH 7.4 containing 1% NP40;
    • 3 washes with 1 ml PBS PH 7.4 containing 0.5 M LiCl;
    • 3 washes with 1 ml 20 mM Tris pH 8, 137 mM NaCl.

    All procedures should be done at 4 °C. However, following the last wash, transfer the tubes to a bench and perform the remainder of the assay for 3-kinase activity at room temperature.

6. Remove as much as possible of the final wash and add the following to each tube:

    • 50 μl of 20 mM Tris pH 8/137 mM NaCl;
    • 10 μl (20 μg) of a PtdIns preparation, made by drying an appropriate amount of PtdIns (solution of 10 mg/ml in chloroform) under an

---

Protocol 2. *Continue*

argon stream. Add 10 mM Tris pH 7.4, 1 mM EGTA to bring the lipid to a concentration of 2 mg/ml and make a liposome suspension by sonicating in an ice bath for 10 min;

- 10 μl of 100 mM MgCl$_2$.

7. Begin the reaction by adding to each tube 5 μl of 0.88 mM ATP + 20 mM MgCl$_2$ + 20 μCi [γ^{32}-P]ATP. Allow the reaction to continue for 15 min with constant agitation.

8. Stop the reaction by adding 20 μl of 6 M HCl and add 160 μl of CHCl$_2$: MeOH (1:1) to the extract. Agitate for an additional 10 min.

9. Centrifuge at 10 000 × *g* for 10 min to separate the two phases.

10. The lower phase (organic) contains the inositol lipids and the upper (aqueous) phase contains the unincorporated [^{32}P]ATP. Pipet 50 μl of the lower phase being careful not to remove any of the upper phase and spot it on an oxalate-treated TLC plate (see above). Resolve the lipids by chromatography in CHCl$_3$:MeOH:H$_2$O:NH$_3$ (60:47:11:3:2), dry the plate completely, and expose for autoradiography.

3. Summary and perspectives

The suggested involvement of PI3K in so many biological processes has provoked much interest in defining the role of this enzyme activity in various signalling pathways. These intense investigations of PI3K emphasize the need for accurate and appropriate assays to measure its lipid kinase activity, as regulated by an expanding collection of ligands. Therefore, we have presented a discussion of both an *in vivo* and *in vitro* assay for measuring PI3K lipid products, in the hope of providing readers with not only methodology but also with the relative advantages of each assay. However, while these assays are useful in obtaining certain information about the activity of PI3K, they are not optimal for fully exploring and appreciating the contributions of lipid products in mediating cell responses. One limitation with current assays is the inability to measure changes in PI3K activity occuring within micro-environments of the cell. PI3K activity is proposed to participate in cellular activities ranging from regulated secretion at the cell's plasma membrane to proliferation, presumbably involving interaction with the cell nucleus. Consequently, if it were possible to localize and quantify the activity of PI3K in such subcellular environments as secretory vesicles or the plasma membrane of intact cells this would be expected to yield valuable insight regarding the role the PI3K in mediating cellular signals. Such *in vivo* measurements might be facilitated by antibodies to lipid products or by development of a membrane permeable indicator that would fluoresce following addition of a phosphate to the lipid, a reagent analagous to fluorescent Ca^{+2} indicators such as fura2. Finally, while this review has focused on the PI3K-generated lipid products and their role in cell signalling, it should be noted that PI3K has two other functions: serine kinase activity, and 'adaptor-like' functions involving the p85 subunit which directs protein-protein interactions with other signalling components. Taken together these various functions of PI3K provide this heteromeric enzyme with the potential to participate in many critical cellular pathways.

References

1. Kapeller, R. and Cantley, L.C., Jr. (1994). *Bio Essays*, **16**, 565–576.
2. Whitman, M., Downes, C.P., Keeler, M., Keller, T., and Cantley, L.C. (1988). *Nature*, **332**, 644–646.
3. Myers, M.G., Jr., Backer, J.M., Sun, X.J., Shoelson, S.E., Hu, P., Schlessinger, J., Yoakim, M., Schaffhausen, B., and White, M.F. (1992). *Proc. Natl. Acad. Sci. USA*, **89**, 10350–10354.
4. Myers, M.G., Jr. and White, M.F. (1993). *Diabetes*, **42**, 643–650.
5. Myers, M.G., Jr., Grammer, T.C., Wang, L.M., Sun, X.J., Pierce, J.H., Blenis, J., and White, M.F. (1994). *J. Biol. Chem.*, **269**, 28783–28789.
6. Pons, S., Asano, T., Glasheen, E.M., Miralpeix, M., Weiland, A., Zhang, Y., Myers, M.G., Jr., Sun, X.J., and White, M.F. (1995). *Mol. Cell Biol.*, **15**, 4453–4465.
7. Inukai, K., Anai, M., van Breda, E., Hosaka, T., Katagiri, H., Funaki, M., Fukushima, Y., Ogihara, T., Yazaki, Y., Kikuchi, M., Oka, Y., and Asano T. (1996). *J. Biol. Chem.*, **271**, 5317–5320.
8. Hiles, I.D., Otsu, M., Volinna, S., Fry, M.J., Gout, I., Dhand, R., Panayotou, G., Ruiz-Larrea, F., Thompson, A., Totty, N.F., Hsuan, J.J., Courtneidge, S.A., Parker P.J., and Waterfield, M.D. (1992). *Cell*, **70**, 419–429.
9. Hu, P., Mondino, A., Skolnik, E.Y., and Schlessinger, J. (1993). *Mol. Cell Biol.*, **13**, 7677–7687.
10. Stoyanov, S., Volinia, S., Hanck, T., Rubio, I., Loubtchenkov, M., Malek, D., Stoyanova, S., Vanhoesenbroek, B., Dhand, R., and Nurnberg, B. (1995). *Science*, **269**, 690–693.
11. Valius, M. and Kazlauskas, A. (1993). *Cell*, **73**, 321–334.
12. Kimura, K., Hattori, S., Kabuyama, Y., Shizawa, Y., Takayanagi, J., Nakamura, S., Toki, S., Matsuda, Y., Onodera, K., and Fukui, Y. (1994). *J. Biol. Chem.*, **269**, 18961–18967.
13. Wennstrom, S., Hawkins, P., Cooke, F., Hara, K., Yonezawa, K., Kasuga, M., Jackson, T., Claessonwelsh, L., and Stephens, L. (1994). *Current Biology*, **4**, 385–393.
14. Yano, H., Nakanishi, S., Kimura, K., Hanai, N., Saitoh, Y., Fukui, Y., Nonomura, Y., and Matsuda, Y. (1993). *J. Biol. Chem.*, **268**, 25846–25856.
15. Yao, R. and Cooper, G.M. (1995). *Science*, **267**, 2003–2006.
16. Chung, J., Grammer, T.C., Lemon, K.P., Kazlauskas, A., and Blenis, J. (1994). *Nature*, **370**, 71–75.
17. Cheatham, R.B., Vlahos, C.J., Cheatham, L., Wang, L., Blenis, J., and Kahn, C.R. (1994). *Mol. Cell Biol.*, **14**, 4902–4911.

18. Rodriguez-Viciana, P., Warne, P.H., Dhand, R., Vanhaesebroeck, B., Gout, I., Fry, M.J., Waterfield, M.D., and Downward, J. (1994). *Nature*, **370**, 527–532.
19. Wood, K.W., Sarnecki, C., Roberts, T.M., and Blenis, J. (1992). *Cell*, **68**, 1041–1050.
20. Hara, K., Yonezawa, K., Sakaue, H., Ando, A., Kotani, K., Kitamura, T., Kitamura, Y., Ueda, H., Stephens, L., Jackson, T.R., Hawkins, P.T., Dhand, R., Clark, A.E., Holman, G.D., Waterfield, M.D., and Kasuga, M. (1994). *Proc. Natl. Acad. Sci. USA*, **91**, 7415–7419.
21. Woscholski, R., Dhand, R., Fry, M.J., Waterfield, M.D., and Parker, P.J. (1994). *J. Biol. Chem.*, **269**, 25 067–25 072.
22. Woscholski, R., Kodaki, T., McKinnon, M., Waterfield, M.D., and Parker, P.J. (1994). *FEBS Lett.*, **342**, 109–114.
23. Nakanishi, S., Catt, K.J., and Balla, T. (1995). *Proc. Natl. Acad. Sci. USA*, **92**, 5317–5321.
24. Cross, M.J., Stewart, A., Hodgkin, M.N., Kerr, D.J., and Wakelam, M.J. (1995). *J. Biol. Chem.*, **270**, 25 352–25 355.
25. Pleiman, C.M., Hertz, W.M., Cambier, J.C. (1994). *Science*, **263**, 1609–1612.
26. Backer, J.M., Myers, M.G., Jr., Shoelson, S.E., Chin, D.J., Sun, X.J., Miralpeix, M., Hu, P., Margolis, B., Skolnik, E.Y., Schlessinger, J., and White, M.F. (1992). *EMBO J.*, **11**, 3469–3479.

A1

Suppliers

Alpha Laboratories Ltd., 40 Parham Drive, Eastleigh, Hants SO5 4NU, UK.

Amicon, Inc., 72 Cherry Hill Drive Beverly, MA 01915, USA.

Avanti Polar Lipids, 700 Industrial Park Drive, Alabaster, AL 35007, USA.

Bellco Glass Inc., 340 Edrudo Road, P.O. Box B, Vineland, NJ 08360-0017, USA; available within the UK from V.A. Howe & Co. Ltd., Beaumont Close, Banbury, Oxon, OX16 7RG, UK.

Bibby Sterilin Ltd., Stone, Staffordshire ST15 0SA, UK.

BioRad Laboratories, Division Headquarters, 3300 Regatta Boulevard, Richmond CA 94804, USA.

Boehringer-Mannheim Corporation, Biochemical Products, 9115 Hague Road, P.O. Box 504 Indianapolis, IN 46250-0414, USA.

BioSoft, P.O. Box 10938, Ferguson, MO 63135, USA.

Branson Ultrasonics Corporation, Eagle Road, Danbury, CT 06810, USA.

Calbiochem-Novachem International, 10394 Pacific Center Court, San Diego, CA 92121, USA.

Cole-Parmer Instrument Corp., 7425 North Oak Park Avenue, Niles, IL 60714, USA

Dupont-NEN Research Products, 549 Albany Street, Boston, MA 02118, USA.

Endecotts Ltd., 9 Lombard Rd., London SW19 3UP, UK.

GIBCO BRL (Life Technologies Inc.), 3175 Staler Read, Grand Island, N.Y. 14072, USA.

Hoefer Scientific Instruments, 654 Minnesota St., San Francisco, CA 94107-3027, USA.

Hoefer Scientific Instruments, UK, Newcastle-Under-Lyme, Staffordshire, UK.

International Equipment Corp., 300 Second Avenue Needham Heights, MA 02194, USA.

Invitrogen Corporation, 3985 B Sorrento Valley Building, San Diego, CA 92121, USA.

J.R.H. Bioscience, 13804 W. 107th Street, Lenexa, KS 66215, USA.

Kontes Glass, 1022 Spruce Street, P.O. Box 729, Vineland, NJ 08360, USA.

Luckhams Ltd., Victoria Gardens, Burgess Hill, Sussex, UK.

Macarthys Surgical Ltd., Dagenham, Essex PMQ 1AD, UK.

Miles Labs, Bayer Corp., Diagnostic Division, 195 West Bird Street, Kankakee, IL 60901, USA.

MRA International, 696 12th Avenue N.E., Naples, FL 33964, USA.

Nunc, A/S Nunc, Kamstrupvej 90, Kamstrup, DK-4000, Roskilde, Denmark.

Parr Instrument Company, 211 53rd Street, Moline, IL 61265, USA; Parr Instruments available within UK from Scientific & Medical Products Ltd, Shirley Institute, 856 Wilmslow Road, Didsbury, Manchester, M20 25A, UK.

Phase Separations Ltd., Deeside Industrial Park, Deeside, Clywd CH5 2NU, UK.

Pierce, P.O. Box 117, Rockford, IL 61105, USA.

Qiagen Inc., 9259 Eton Avenue, Chatsworth, CA 91311, USA.

Richardsons of Leicester Ltd., Evington Valley Road, Leicester LE5 5LJ, UK.

Russell pH Ltd., Station Road, Auchtermuchty, Fife KY14 7DP, Scotland, UK.

Santa Cruz Biotechnology, Inc. 2161 Delaware Ave., Santa Cruz, CA 95060, USA.

Savant Instruments, Inc., 110-103 Bi-County Boulevard, Framingdale, NY 11735, USA.

Sigma Chemical Company, 3050 Spruce Street, P.O. Box 14508, St Louis MO 63178, USA.

Spectrum Medical Industries, 1100 Rankin Rd. Huston, TX 77073-4716, USA.

Takara Shuzo Co. Ltd., Biomedical Group 3-4-1 Seta, Otsu, Shiga 520-21, Japan.

Tosoh Corporation, 3-2-4 Kyobashi, Chuoh-ku, Tokyo 104, Japan.

Upstate Biochemical Incorporated, 199 Saranac Ave., Lake Placid, NY 12946, USA.

Virtis Company, 815 Route 208, Gardiner, NY 12525, USA.

Wallac UK, 20 Vincent Avenue, Crownhill, Milton Keynes MK8 0AB, UK.

Whatman International Ltd., Whatman House, St. Leonards Road, Maidstone, Kent ME16 0LS, UK.

Whatman Lan Sales, P.O. Box 1359 Hillsboro, OR 97123, USA.

Yakult Pharmaceutical Co. Ltd., Nishinomya, Japan.

A2

Suppliers of inositol derivatives

Research into inositol derivatives is frequently dependent upon many compounds that are not commercially available. This appendix contains a list of a variety of inositol-based compounds. The individuals associated with each compound have indicated their willingness in principle to supply small amounts of these derivatives without charge, purely for research purposes in academic laboratories (for a compound to be included in this list, the inositol moiety had to be present). It should be noted that each potential supplier reserves the right not to agree to a request to provide a particular compound (for example, if a conflict of interest arises). Furthermore, it should always be assumed that such compounds are offered on a collaborative basis unless a different arrangement is explicitly negotiated. In other words, a supplier and his/her colleagues should be considered to be entitled, if they so wish, to co-authorship of any ensuing publication resulting from the use of their materials.

In the list of structures, inositol can be assumed to be the *myo*-epimer unless otherwise indicated. Substituents around the inositol ring are numbered with reference to the D-1 nomenclature adopted by IUPAC (see *Biochem. J.* (1989) **258** 1–2), unless the designation is 'D/L', which indicates a racemic mixture. Many of these compounds will contain counter-ions; these are not indicated here for reasons of brevity. Each potential recipient is advised to determine that there are no inacurracies in the description of these compounds, and they should also ascertain the optimum conditions for use and storage. Some of the compounds are accompanied by a laboratory's suggested potential application, but this is meant as a guide only, and is not designed to be an exclusive list.

Inositol derivatives

Inositol analogues

3-Azido-3-deoxy-inositol **[11]**
   A weak substrate of PtdIns 3-kinase. A potent inhibitor of the growth of 3T3 cells.
3-Chloro-3-deoxy-inositol **[11]**

Inositol monophosphates

Inositol 1-monophosphate [6]
D/L-Inositol 1-monophosphate [4]
Inositol 1,2-cyclic phosphate [6]
Inositol 2-monophosphate [4]
D/L-Inositol 4-monophosphate [4]
Inositol 5-monophosphate [4]

Inositol bisphosphates

D/L-Inositol 1,2-bisphosphate [4,7]
Inositol 1,3-bisphosphate [4]
Inositol 1,4-bisphosphate [13]
   Activator of human DNA polymerase alpha
D/L-Inositol 1,4-bisphosphate [4]
D/L-Inositol 1,5-bisphosphate [4]
D/L-Inositol 1,6-bisphosphate [4]
D/L-Inositol 2,4-bisphosphate[4]
Inositol 2,5-bisphosphate [4]
Inositol 2,6-bisphosphate [4]
Inositol 4,5-bisphosphate [13]
D/L-Inositol 4,5-bisphosphate [4]
Inositol 4,6-bisphosphate [4]

Inositol trisphosphates

Inositol 1,2,3-trisphosphate [4,7]
   Natural compound, found in high concentrations: It chelates iron and shows anti-oxidant properties (which may be linked with its biological function), and it is a siderophore, catalysing iron uptake into *Pseudomonas aeruginosa*.
D/L-Inositol 1,2,4-trisphosphate [4]
D/L-Inositol 1,2,5-trisphosphate [4]
Inositol 1,2,6-trisphosphate [4,6]
D/L-Inositol 1,2,6-trisphosphate [4]
Inositol 1,3,4-trisphosphate [3,6,8,13]
D/L-Inositol 1,3,4-trisphosphate [4]
Inositol 1,3,5-trisphosphate [4]
Inositol 1,3,6-trisphosphate [8]
   Potent calcium-mobilizing isomer
Inositol 1,4,5-trisphosphate [5,6,13]
   Calcium mobilizing second messenger.
D/L-Inositol 1,4,5,-trisphosphate [4]
Inositol 1,4,6-trisphosphate [3,8]
D/L-Inositol 1,4,6-trisphosphate [4]

Inositol 1,5,6-trisphosphate [13]
D/L-Inositol 1,5,6-trisphosphate [4]
D/L-Inositol 2,4,5-trisphosphate [3,4]
Inositol 2,4,6-trisphosphate [4]
Inositol 3,4,5-trisphosphate [2]
   Constituent of rat mammary tumor cells and avian erythrocytes.
Inositol 3,4,6-trisphosphate [3]
Inositol 3,4,6-trisphosphate [3,8]
Inositol 3,5,6-trisphosphate [6]
   Enantiomer of Ins$(1,4,5)P_3$
Inositol 4,5,6-trisphosphate [4]

Inositol trisphosphate analogues

2-O-(p-aminobenzoyl)-inositol 1,4,5-trisphosphate [3]
D/L-2-O-(p-azidobenzoyl)-inositol 1,4,5-trisphosphate [3]
2-O-(p-aminocyclohexanecarbonyl)-inositol 1,4,5-trisphosphate [3]
D/L-3-O-(m-aminobenzoyl)-inositol 1,4,5-trisphosphate [3]
D/L-3-O-(p-aminobenzoyl)-inositol 1,4,5-trisphosphate [3]
D/L-3-O-(p-benzoyl)-inositol 1,4,5-trisphosphate [3]
3-Amino-3-deoxy-inositol 1,4,5-trisphosphate [11]
   A pH-dependent partial Ins$(1,4,5)P_3$-agonist.
D/L-di-2,6-O-butyryl-3-O-methyl-inositol 1,4,5-trisphosphate [1]
3-Chloro-3-deoxy-inositol 1,4,5-trisphosphate [11]
   A relatively metabolically stable 3-position blocked full Ins$(1,4,5)P_3$-agonist.
3-Deoxy-inositol 1,4,5-trisphosphate [11]
   A 3-position blocked full Ins$(1,4,5)P_3$-agonist.
D/L-3-O-(p-fluorobenzoyl)-inositol 1,4,5-trisphosphate [3]
3-Fluoro-3-deoxy-myo-inositol 1,4-bisphosphate-5-phosphorothioate [11]
   A full Ins$(1,4,5)P_3$-agonist, less potent than Ins$(1,4,5)P_3$ but resistant to 3-kinase and 5-phosphatase.
3-Fluoro-3-deoxy-myo-inositol 1,5-bisphosphate-4-phosphorothioate [11]
   A full Ins$(1,4,5)P_3$-agonist, less potent than Ins$(1,4,5)P_3$ but resistant to 3-kinase and 5-phosphatase.
3-Fluoro-3-deoxy-inositol 1,4,5-trisphosphate [11]
   A relatively metabolically stable 3-position blocked full Ins$(1,4,5)P_3$-agonist.
D/L-3-O-(p-hydroxybenzoyl)-inositol 1,4,5-trisphosphate [3]
D/L-3-O-(p-methoxybenzoyl)-inositol 1,4,5-trisphosphate [3]
2,3-(Methoxymethylene)-6-(4,5-dimethoxy-2-nitrobenzyl)-inositol 1,4,5-trisphosphate hexakis(propionyloxymethyl) ester [9]
   A membrane-permeant ester of caged Ins$(1,4,5)P_3$ that loads cells with 2,3-(methoxymethylene)-6-(4,5-dimethoxy-2-nitrobenzyl)-inositol 1,4,5-trisphosphate, which is trapped, biologically inert, and metabolically stable

inside cells until photolysed. Photolysis releases 2,3-(methoxymethylene)-inositol 1,4,5-trisphosphate; the latter cannot be phosphorylated at the 3-position, but it binds to and activates the Ins(1,4,5)P_3 receptor with approximately the same affinity as Ins(1,4,5)P_3 itself.

D/L-3-*O*-(*p*-methylbenzoyl)-inositol 1,4,5-trisphosphate [3]

5-Deoxy-5-methylenephosphonyl-inositol 1,4-bisphosphate [2]
  A poorly metabolized analogue of Ins(1,4,5)P_3.

L-*chiro*-Inositol 1,2,3-trisphosphate [4]

scyllo-Inositol 1,2,4-trisphosphate [8]
  Potent calcium mobilizer.

Inositol 1,3,5-trisphosphorothioate [8]
  Ins(1,4,5)P_3/Ins(1,3,4,5)P_4 5-phosphatase inhibitor.

Inositol 1,3,4-trisphosphate with a P-1 aminopropyl diester linkage to either Affigel, or to photoaffinity, biotinylated or fluorescent probes [10]

Inositol 1,3,6-trisphosphorothioate [8]
  Partial Ins(1,4,5)P_3 antagonist.

Inositol 1,4,5-trisphosphate, with a P-1 aminopropyl diester linkage to either Affigel, or to photoaffinity, biotinylated or fluorescent probes [10]
  For identification of receptors/binding sites for Ins(1,4,5)P_3 and PtdIns(4,5)P_2.

D/L-Inositol 1,4,5-trisphosphate hexakis(propionyloxymethyl) ester [9]
  A membrane-permeant ester of Ins(1,4,5)P_3 that releases intracellular Ca^{2+} when applied extracellularly at concentrations of 10–100 mM. Its ability to elevate [Ca^{2+}]$_i$ is blocked by intracellular heparin and is potentiated by thimerosal. Orthophosphate tris(propionylmethyl) ester [9] is the control for byproducts of propionyloxymethyl ester hydrolysis; this usually has no effect on [Ca^{2+}]$_i$ at extracellular concentrations up to several hundred mM.

D/L-Inositol 1,4,5-trisphosphate hexakis(butyryloxymethyl) ester [9]
  A membrane-permeant ester of Ins(1,4,5)P_3 that releases intracellular Ca^{2+} when applied extracellularly at concentrations of 1–10 mM, though its maximum effect is less than that of the propionyl derivative.

Inositol 1,4,6-trisphosphorothioate [8]
  Partial Ins(1,4,5)P_3 antagonist.

L-*chiro*-inositol 1,4,6-trisphosphothioate [8]
  Ins(1,4,5)P_3/Ins(1,3,4,5)P_4 5-phosphatase inhibitor

L-*chiro*-inositol 2,3,5-trisphosphate [8]
  Ins(1,4,5)P_3 3-kinase inhibitor.

L-*chiro*-inositol 2,3,5-trisphosphothioate [8]
  Inhibitor of Ins(1,4,5)P_3/Ins(1,3,4,5)P_4 5-phosphatase, Ins(1,4,5)P_3 3-kinase and PtdIns 3-kinase.

Inositol 3,5,6-trisphosphorothioate [8]
  Ins(1,4,5)P_3 / Ins(1,3,4,5)P_4 5-phosphatase inhibitor.

D/L-3-O-methyl-inositol 1,4,5-trisphosphate [1]

A potent agonist for the activation of calcium release from internal stores which cannot be phosphorylated to Ins(1,3,4,5)P_4.

Inositol tetrakisphosphates

D/L-inositol 1,2,3,4-tetrakisphosphate [4]
Inositol 1,2,3,5-tetrakisphosphate [4]
Inositol 1,2,4,5-tetrakisphosphate [4,8]
  Potent calcium-mobilizing InsP_4.
D/L-inositol 1,2,4,5-tetrakisphosphate [3,4]
D/L-inositol 1,2,4,6-tetrakisphosphate [4]
Inositol 1,2,5,6-tetrakisphosphate [4,13]
D/L-inositol 1,2,5,6-tetrakisphosphate [4]
Inositol 1,3,4,5-tetrakisphosphate [2,5,8,13]
  Calcium mobilizing second messenger.
D/L-inositol 1,3,4,5-tetrakisphosphate [4]
Inositol 1,3,4,6-tetrakisphosphate [3,4,13]
Inositol 1,3,5,6-tetrakisphosphate [8]
Inositol 1,4,5,6-tetrakisphosphate [3,6,13]
D/L-inositol 1,4,5,6-tetrakisphosphate [4]
Inositol 2,3,5,6-tetrakisphosphate [8]
Inositol 2,4,5,6-tetrakisphosphate [4]
Inositol 3,4,5,6-tetrakisphosphate [1,3,6,13]
  Intracellular second messenger; inhibitor of calcium-activated chloride channels.

Inositol tetrakisphosphate analogues

D/L-2-O-butyryl-1-O-methyl-inositol 3,4,5,6-tetrakisphosphate [1]
2-deoxy-inositol 1,3,4,5-tetrakisphosphate [8]
1,2-di-O-butyryl-inositol 3,4,5,6-tetrakisphosphate [1]
1,2-di-O-butyryl-inositol 3,4,5,6-tetrakisphosphate-octakis(acetoxymethyl) ester [1]
  A membrane-permeant derivative of Ins(3,4,5,6)P_4, that has been shown to inhibit calcium-activated chloride secretion in monolayers of T84 cells.
D/L-1,2-di-chloro-inositol 3,4,5,6-tetrakisphosphate [1]
1,2-di-O-methyl-inositol 3,4,5,6-tetrakisphosphate [1]
D/L-1,2-dideoxy-1,2-difluoro-myo-inositol 3,4,5,6-tetrakisphosphate [7]
  Has the potential of being a sustrate/inhibitor of enzymes, or agonist/ antagonist of receptors that utilize Ins(3,4,5,6)P_4 or its enantiomer, Ins(1,4,5,6)P_4.
D/L-1,2-dideoxy-1,2-difluoro-*scyllo*-inositol 3,4,5,6-tetrakisphosphate [7]
  Has the potential of being a sustrate/inhibitor of enzymes, or agonist/ antagonist of receptors that utilize Ins(3,4,5,6)P_4 or its enantiomer, Ins(1,4,5,6)P_4.

2-fluoro-scyllo-inositol 3,4,5,6-tetrakisphosphate [1]
3-fluoro-3-deoxy-inositol 1,2,4,5-tetrakisphosphate [11]
  A 2,3-position blocked full Ins(1,4,5)P_3-agonist.
L-chiro-Inositol 1,2,3,5-tetrakisphosphate [4]
Inositol 1,2,4,5-tetrakisphosphate with a P-1 aminopropyl diester linkage to either Affigel, or to photoaffinity, biotinylated or fluorescent probes [10]
Inositol 1,2,5,6-tetrakisphosphate with a P-5 aminopropyl diester linkage to either Affigel, or to photoaffinity, biotinylated or fluorescent probes [10]
Inositol 1,3,4,5-tetrakisphosphate with a P-1 aminopropyl diester linkage to either Affigel, or to photoaffinity, biotinylated or fluorescent probes [10]
  For purification and active site labeling of Ins(1,3,4,5)P_4 binding domains on coatomer subunits, AP-2, AP-3 and synaptotagmin; for labeling and purification of centaurin and other PtdIns(3,4,5)P_3 binding proteins.
Inositol 1,4,5-trisphosphate-3-phosphorothioate [11]
  A 3-phosphatase resistant analogue of Ins(1,3,4,5)P_4.
scyllo-inositol 1,2,3,5-tetrakisphosphate [8]
scyllo-inositol 1,2,4,5-tetrakisphosphate [8]
3-O-methyl-inositol 1,4,5,6-tetrakisphosphate [1]
1-O-methyl-inositol 3,4,5,6-tetrakisphosphate [1]
D/L-*scyllo*-inositol 3,4,5,6-tetrakisphosphate [1]

Inositol pentakisphosphates

Inositol 1,2,3,4,5-pentakisphosphate [1]
D/L-Inositol 1,2,3,4,5-pentakisphosphate [4]
Inositol 1,2,3,4,6-pentakisphosphate [1,4]
Inositol 1,2,3,5,6-pentakisphosphate [1]
Inositol 1,2,4,5,6-pentakisphosphate [1,3]
D/L-Inositol 1,2,4,5,6-pentakisphosphate [4]
Inositol 1,3,4,5,6-pentakisphosphate [3,4,6,13]
Inositol 2,3,4,5,6-pentakisphosphate [1]

Inositol hexakisphosphate analogues

P(1,2)- and P(4,5)-(*o*-Nitrobenzyl) esters of InsP_6 [10]
  'Caged' InsP_6.
Inositol hexakisphosphate with a P-2 aminohexyl phosphodiester linkage to either Affigel, or to photoaffinity, biotinylated or fluorescent probes [10]
  For purification and active site labeling of InsP_6 binding domains on coatomer subunits, AP-2, AP-3, and synaptotagmin.
Inositol 1,2,4,5,6-pentakisphosphate 3-diphosphate [2]
Inositol 1,2,3,5,6-pentakisphosphate 4-diphosphate [2]
Inositol 1,2,3,4,6-pentakisphosphate 5-diphosphate [2]
Inositol 1,2,3,4,5-pentakisphosphate 6-diphosphate [2]

Inositol 1,3,4,5,6-pentakisphosphate 2-diphosphate **[2]**
Inositol 2,3,4,5,6-pentakisphosphate 1-diphosphate **[2]**

Inositol lipid analogues

1,2-dibutyl-*sn*-glycero-3-phospho-(1-inositol-2-fluorodeoxy-4,5-bisphosphate) **[12]**

Water-soluble competitive inhibitor of mammalian phospholipase C, stable to non-specific chemical hydrolysis, for co-crystallization with phospholipase C isozymes in X-ray crystallography.

1,2-dibutyl-*sn*-glycero-3-phospho-(1-inositol-3-fluorodeoxy-4,5-bisphosphate) **[12]**

Water-soluble competitive inhibitor of phosphoinositide 3-kinase, stable to non-specific chemical hydrolysis, for co-crystallization with 3-kinase isozymes in X-ray crystallography.

1,2-dibutyryl-*sn*-glycero-3-phospho-(1-inositol-3-bisphosphate) **[10]**
1,2-dibutyryl-*sn*-glycero-3-phospho-(1-inositol-3,4-bisphosphate) **[10]**
1,2-dibutyryl-*sn*-glycero-3-phospho-(1-inositol-3,4,5-bisphosphate) **[10]**
1,2-dihexanoyl-*sn*-glycero-3-phospho-(1-inositol) **[12]**
1,2-dihexanoyl-*sn*-glycero-3-phospho-(1-inositol-4,5-bisphosphate) **[12]**
1,2-dihexanoyl-*sn*-glycero-3-phospho-(1-inositol-3,4,5-trisphosphate) **[12]**
1,2-dioctanoyl-*sn*-glycero-3-phospho-(1-inositol-3-bisphosphate) **[13]**
1,2-dioctanoyl-*sn*-glycero-3-phospho-(1-inositol-3,4-bisphosphate) **[2,10,13]**
1,2-dioctanoyl-sn-glycero-3-phospho-(1-inositol-3,4,5-trisphosphate) heptakis (acetoxymethyl) ester **[9]**

Derivative of PtdIns(3,4,5)P_3 with all negative charges masked as intracellularly hydrolyzable esters, to increase membrane permeability.

1,2-dioctanoyl-*sn*-glycero-3-phospho-(1-inositol-3,4,5-trisphosphate) **[2,10,13]**

Second messenger analogue.

1,2-dioctanoyl-*sn*-glycero-3-phospho-(1-inositol-4,5-bisphosphate) **[10]**
1,2-dioctyl-sn-glycero-3-phospho-(1-inositol-3,4,5-trisphosphate) **[2]**

Water soluble version of PtdIns(3,4,5)P_3 with octyl ethers at *sn*-1 and *sn*-2 positions of glycerol.

1,2-dipalmitoyl-*sn*-glycero-3-phospho-(1-inositol) **[6,12]**
1,2-dipalmitoyl-*sn*-glycero-3-phospho-(1-inositol-2-fluorodeoxy) **[12]**
1,2-dipalmitoyl-*sn*-glycero-3-phospho-(1-inositol-2-fluorodeoxy-4,5-bisphosphate) **[12]**

Competitive inhibitor of mammalian phospholipase C.

1,2-dipalmitoyl-*sn*-glycero-3-phospho-(1-inositol-3-fluorodeoxy-4,5-bisphosphate) **[12]**

Competitive inhibitor of mammalian phospholipase C.

1,2-dipalmitoyl-*sn*-glycero-3-phospho-(1-inositol-2,3,6-trifluorodeoxy-4,5-bisphosphate) **[12]**

Competitive inhibitor of phosphoinositide 3-kinase isozymes.

1,2-dipalmitoyl-*sn*-glycero-3-phospho-(1-inositol-3,6-difluorodeoxy-4,5-bis-phosphate) [12]

Competitive inhibitor of phosphoinositide 3-kinase isozymes.

1,2-dipalmitoyl-*sn*-glycero-3-phospho-(1-inositol-4,5-bisphosphate) [12]

For applications as substrate of PtdIns kinases, and in biophysical measurements of model bilayers, or for crystallization efforts.

1,2-dipalmitoyl-*sn*-glycero-3-thiophospho-1-inositol [6]

Phospholipase C resistant thiono-analogue of PtdIns

2R-1,2-dipalmitoyloxypropyl-3-mercaptophospho-(1-inositol) [6]

Thiol-analogue of PtdIns suitable for spectrophotometric, continuous assay of PtdIns hydrolysis by PLC.

1,2-dipalmitoyl-*sn*-glycero-3-phospho-(1L-2-*chiro*-inositol) [6]

A *chiro*-inositol analogue of PtdIns for metabolic studies.

1,2-dipalmitoyl-*sn*-glycero-3-thiophospho-(1-inositol-3,4,5-tris-phosphorothioate) [6]

Analogue of PtdIns(3,4,5)P_3 with projected long life-time under biological conditions due to resistance to metabolism by PLC, PLD and phosphatases.

1,2-dipalmitoyl-*sn*-glycero-3-phospho-(1-inositol-3,4,5-trisphosphorothioate) [6]

Analogue of PtdIns(3,4,5)P_3 with anticipated long life-time under biological conditions due to resistance to metabolism by PLC and phosphatases.

1,2-dipalmitoyl-*sn*-glycero-3-phospho-(1-inositol-3-phosphate) [6,10,13]

1,2-dipalmitoyl-*sn*-glycero-3-phospho-(1-inositol-3,4-bisphosphate) [6,10,13]

1,2-dipalmitoyl-*sn*-glycero-3-phospho-(1-inositol-3,4,5-trisphosphate) [6,10,12,13]

1,2-dipalmitoyl-*sn*-glycero-3-phospho-(1-inositol-4-phosphate) [10]

2-O-(2-Amino-2-Deoxy-a-D-Glucopyranosyl)-D-*chiro*-Inositol-1-Phosphate [2]

Truncated form of insulin intracellular second messenger.

Biotinylated PtdIns(3,4,5)P_3 [6,10]

Affinity ligand with possible application for purification of enzymes and proteins that interact with PtdIns(3,4,5)P_3.

ω-aminodioctanoyl phosphatidylinositol (3,4,5)-trisphosphate [2]

Water soluble version of PtdIns(3,4,5)P_3 with octyl ethers and amino function at ω-end of *sn*-1 alkyl chain.

1-hexanoyl-2-2-(ω-aminobutanoyl)-sn-glycero-3-phospho-(1-inositol-4,5-bis-phosphate) [12]

The primary amino function in this lipid analogue allows conjugation to photoaffinity, fluorescent and other reporter groups, and to biotin, affinity column or gold surfaces.

P-1 triester affinity analogues of PtdIns(4,5)P_2, PtdIns(3,4)P_2, PtdIns(3,4,5)P_3 all with dipalmitoyl chains [10]

Suppliers

[1] Dr. C. Schultz, Zentrum für Umweltforschung und Technologie, Universität Bremen, UFT, Leobener Str. 28359 Bremen, Germany. Fax: 421-218-4264; Tel: 421-218-7665; e-mail: schultz@chemie.uni-bremen.de.

[2] Dr. J. R. Falck, Department of Molecular Genetics, University of Texas, Southwestern Medical Center, 5323 Harry Hines Blvd., Dallas 75235, USA. Tel: 214-648-3628; fax: 214-648-7539.

[3] Dr. M. Hirata, Department of Biochemistry, Kyushu University, Faculty of Dentistry, 3-1-1 Maidashi Higashi-ku, Fukuoka 812, Japan. Tel: 92-641-1151; fax: 92-631-2731 e-mail mhiraded@mbox.nc.kyushu-u.ac.jp.

[4] Dr. S.K. Chung, Department of Chemistry, Pohang University of Science and Technology, San 31 Hyoja Dong, Pohang, 790-784 Korea. Fax: 562-279-3399; tel: 562-279-2103; e-mail skchung@chem.postech.ac.kr.

[5] Dr. R. Irvine, Dept of Pharmacology, Cambridge University, Tennis Court Road, Cambridge, CB2 3ES UK. Tel: 1223-339-683; fax: 1223-334-040; e-mail rfi20@cam.ac.uk.

[6] Dr. K. S. Bruzik, University of Illinois at Chicago, Department of Medicinal Chemistry and Pharmacognosy (M/C 781), College of Pharmacy, 833 South Wood Street, Chicago, Illinois 60612-7231, USA. Tel: 312-996-4576; fax: 312-996-7107; e-mail: u63800@uicvm.uic.edu.

[7] Dr. S. Freeman, Department of Pharmacy, University of Manchester, Oxford Road, Manchester, M13 9PL, UK. Tel: 161-275-2366; fax: 161-275-2396; e-mail: sfreeman@fs1.pa.man.ac.uk.

[8] Dr. B.V.L. Potter, University of Bath, Department of Medicinal Chemistry, School of Pharmacy and Pharmacology, Claverton Down, Bath BA2 7AY UK. Tel: 1225-826-639; fax: 1225-826-114; e-mail prsbvlp@bath.ac.uk.

[9] Dr. R. Y. Tsien, Howard Hughes Medical Institute 0647, 310 Cellular and Molecular Medicine West, University of California, San Diego, 9500 Gilman Drive, La Jolla, CA 92093-0647, USA. Tel: 619-534-4891; fax: 619-534-5270; e-mail rtsien@ucsd.edu.

[10] Dr. G.D. Prestwich, Department of Medicinal Chemistry, University of Utah, Salt Lake City Utah, UT 84112, USA. Tel: 801-585-9051; fax: 801-585-9053; e-mail: gprestwich@deans.pharm.utah.edu.

[11] Dr. A. H. Fauq, Mayo Clinic Jacksonville, 4500 San Pablo Road, Jacksonville, Florida 32224, USA. Tel: 904-953-2000.

[12] Dr. R. Aneja, Nutrimed Biotech, Cornell University Research Park, 270-276 Langmuir Laboratory, Ithaca, NY 14850, USA. Tel:/Fax: 607-257-1166.

[13] Dr. C.-S. Chen, Division of Medicinal Chemistry and Pharmaceutics, College of Pharmacy, University of Kentucky, KY 40536-0082, USA. Tel: 606-257-2300; fax: 606-257-2489; e-mail: cchen1@pop.uky.edu.

Index